18th Edition
IET Wiring Regulations
Wiring Systems and Fault Finding
for Installation Electricians

This book deals with an area of practice that many students and non-electricians find particularly challenging. It explains how to interpret circuit diagrams and wiring systems, and outlines the principles of testing before explaining how to apply this knowledge to fault finding in electrical circuits.

A handy pocket guide for anybody who needs to be able to trace faults in circuits, whether in domestic, commercial or industrial settings, this book will be extremely useful to electricians, plumbers, heating engineers and intruder alarm installers.

- Fully up to date with the 18th Edition IET Wiring Regulations 2018.
- Covers all the principles and practice of testing and fault diagnosis in a way that is clear for students and non-electricians.
- Expert advice from an engineering training consultant, supported with colour diagrams and key data.

Brian Scaddan, I Eng, MIET, is an Honorary Member of City & Guilds and has over 45 years' experience in further education and training. He was Director of Brian Scaddan Associates Ltd, an approved training centre offering courses on all aspects of electrical installation contracting, including those for City & Guilds and EAL. He is also a leading author of books on other installation topics.

By the same author

18th Edition IET Wiring Regulations: Design and Verification of Electrical Installations, 9th ed, 978-1-138-60600-5

18th Edition IET Wiring Regulations: Explained and Illustrated, 11th ed, 978-1-138-60605-0

18th Edition IET Wiring Regulations: Electric Wiring for Domestic Installers, 16th ed, 978-1-138-60602-9

18th Edition IET Wiring Regulations: Inspection, Testing and Certification, 9th ed, 978-1-138-60607-4

Electrical Installation Work, 8th ed, 978-1-138-84927-3

PAT: Portable Appliance Testing, 4th ed, 978-1-138-84929-7

The Dictionary of Electrical Installation Work, 978-0-08-096937-4

18th Edition
IET Wiring Regulations
Wiring Systems and Fault Finding
for Installation Electricians

7th Edition

Brian Scaddan

Routledge
Taylor & Francis Group
LONDON AND NEW YORK

Seventh edition published 2019
by Routledge
2 Park Square, Milton Park, Abingdon, Oxon, OX14 4RN

and by Routledge
711 Third Avenue, New York, NY 10017

Routledge is an imprint of the Taylor & Francis Group, an informa business

First edition published 1991 by Newnes, an imprint of Elsevier
Sixth edition published by Routledge 2015

British Library Cataloguing-in-Publication Data
A catalogue record for this book is available from the British Library

Library of Congress Cataloging-in-Publication Data
A catalog record has been requested for this book

ISBN: 978-1-138-60611-1 (hbk)
ISBN: 978-1-138-60609-8 (pbk)
ISBN: 978-0-429-46687-8 (ebk)

Typeset in Kuenstler 480 and Trade Gothic by
Servis Filmsetting Ltd, Stockport, Cheshire

Contents

Preface

The aim of this book is to help the reader to approach the drawing and interpretation of electrical diagrams with confidence, to understand the principles of testing and to apply this knowledge to fault finding in electrical circuits.

The abundant colour diagrams with associated comments and explanations lead from the basic symbols and simple circuit and wiring diagrams, through more complex circuitry, to specific types of wiring systems and, finally, to the methodical approach to fault finding.

The new edition has been brought fully in line with the 18th Edition IET Wiring Regulations.

Brian Scaddan

CHAPTER 1

Diagrams

Important terms/topics covered in this chapter:

- BS EN 60617 symbols
- Diagrams
- Circuit convention
- Relay logic

By the end of this chapter the reader should:

- be aware of the correct symbols to be used on diagrams,
- know the different types of diagrams in general use and why they are used,
- understand circuit convention and its importance in interpreting diagrams,
- understand simple relay logic and its application to PLCs.

This is an area often overlooked or even ignored. The IET Wiring Regulations require that 'diagrams, charts, tables or equivalent forms of information are made available' to the installer and inspector and tester.

BS EN 60617 SYMBOLS

BS EN 60617 gives the graphical symbols that should be used in all electrical/electronic diagrams or drawings. Since the symbols fall in line with the International Electrotechnical Commission (IEC) document 617, it should be possible to interpret non-UK diagrams. Samples of the symbols used in this book are shown in Figure 1.1.

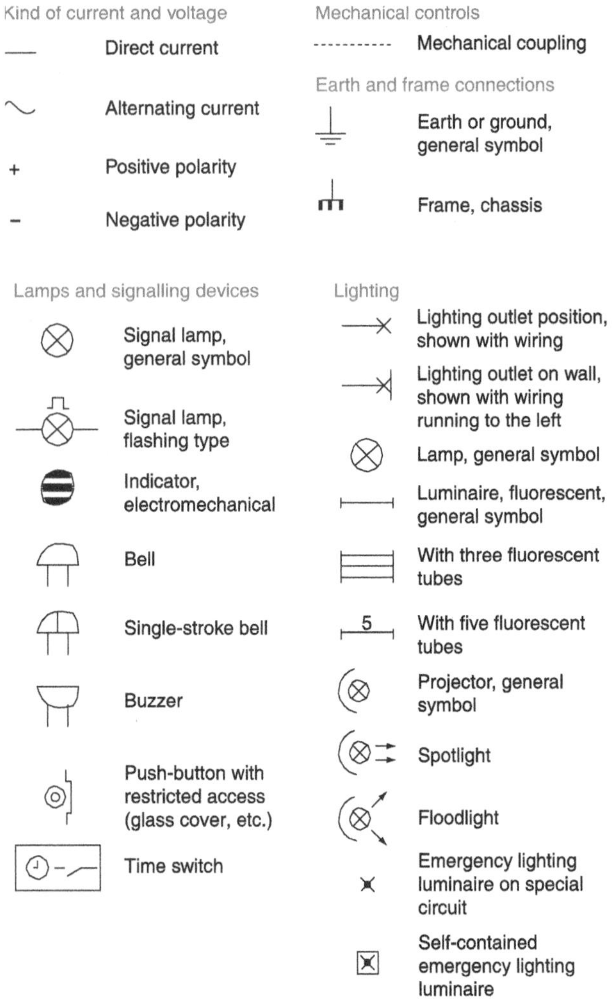

Kind of current and voltage

——	Direct current
∿	Alternating current
+	Positive polarity
–	Negative polarity

Lamps and signalling devices

⊗	Signal lamp, general symbol
–⊗–	Signal lamp, flashing type
⊜	Indicator, electromechanical
⊓	Bell
⊕	Single-stroke bell
⊻	Buzzer
⊚	Push-button with restricted access (glass cover, etc.)
⊙–⁄	Time switch

Mechanical controls

- - - - - -	Mechanical coupling

Earth and frame connections

⏚	Earth or ground, general symbol
rⱈ	Frame, chassis

Lighting

——✕	Lighting outlet position, shown with wiring
——✕∣	Lighting outlet on wall, shown with wiring running to the left
⊗	Lamp, general symbol
∣——∣	Luminaire, fluorescent, general symbol
≣	With three fluorescent tubes
—5—	With five fluorescent tubes
(⊗	Projector, general symbol
(⊗⇉	Spotlight
(⊗	Floodlight
✕	Emergency lighting luminaire on special circuit
✕	Self-contained emergency lighting luminaire

FIGURE 1.1   BS EN 60617 symbols.

Miscellaneous

 Antenna

 Fan, shown with wiring

 Distribution centre, shown with five conduits

 Intercommunication instrument

 Water heater, shown with wiring

Architectural and topographical installation plans and diagrams

Socket outlets

 Socket outlet (power), general symbol

 3

 Three outlets shown: two forms

 With single-pole switch

 Socket outlet (power) with isolating transformer, for example shaver outlet

Socket outlet (telecommunications), general symbol

Designations are used to distinguish different types of outlets:

TP = Telephone
M = Microphone
⊐ = Loudspeaker
FM = Frequency modulation
TV = Television
TX = Telex

Switches

 Switch, general symbol

 Switch with pilot light

 Switch, two pole

 Two-way switch, single pole

 Intermediate switch

 Dimmer

 Pull-cord switch, single pole

 Push-button

 Push-button with indicator lamp

FIGURE 1.1 *(Continued)*

Switchgear, control gear and protective devices

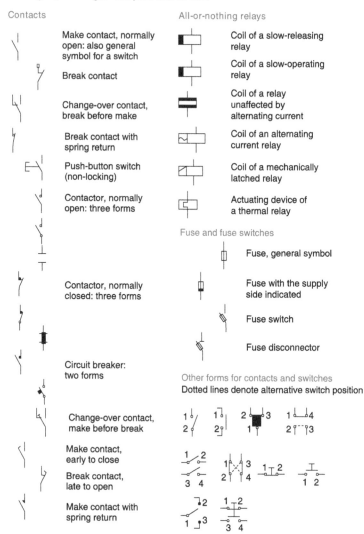

Contacts

Make contact, normally open: also general symbol for a switch

Break contact

Change-over contact, break before make

Break contact with spring return

Push-button switch (non-locking)

Contactor, normally open: three forms

Contactor, normally closed: three forms

Circuit breaker: two forms

Change-over contact, make before break

Make contact, early to close

Break contact, late to open

Make contact with spring return

All-or-nothing relays

Coil of a slow-releasing relay

Coil of a slow-operating relay

Coil of a relay unaffected by alternating current

Coil of an alternating current relay

Coil of a mechanically latched relay

Actuating device of a thermal relay

Fuse and fuse switches

Fuse, general symbol

Fuse with the supply side indicated

Fuse switch

Fuse disconnector

Other forms for contacts and switches
Dotted lines denote alternative switch position

FIGURE 1.1 BS EN 60617 symbols (*Continued*)

DIAGRAMS

The four most commonly used diagrams are the block diagram, inter-connection diagram, the circuit or schematic diagram and the wiring or connection diagram.

Block diagrams

These diagrams indicate, by means of block symbols with suitable notes, the general way in which a system functions. They do not show detailed connections (Figures 1.2a and b).

Interconnection diagrams

In this case, items of equipment may be shown in block form but with details of how the items are connected together (Figure 1.3).

Circuit or schematic diagrams

These diagrams show how a system works, and need to pay no attention to the actual geographical layout of components or parts of components

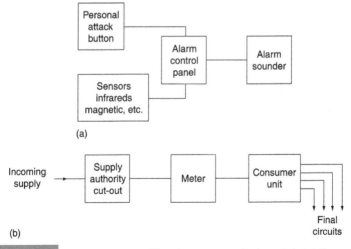

FIGURE 1.2 (a) Security system, (b) Intake arrangement for domestic installation.

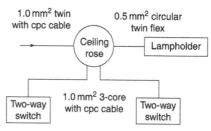

FIGURE 1.3 Two-way lighting system.

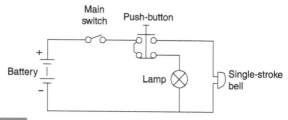

FIGURE 1.4

in that system. For example, a pair of contacts which form part of, say, a timer may appear in a different and quite remote part of the diagram than the timer operating coil that actuates them. In this case some form of cross-reference scheme is needed (e.g. T for the timer coil and T1, T2, T3, etc. for the associated contacts).

It is usual for the sequence of events occurring in a system to be shown on a circuit diagram from left to right or from top to bottom. For example, in Figure 1.4, nothing can operate until the main switch is closed, at which time the signal lamp comes on via the closed contacts of the push-button. When the push is operated the lamp goes out and the bell is energized via the push-button's top pair of contacts.

Wiring or connection diagrams

Here the diagrams show how a circuit is to be actually wired. Whenever possible, especially in the case of control panels, they should show components in their correct geographical locations.

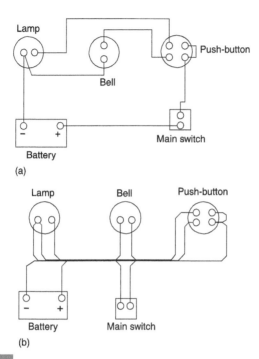

FIGURE 1.5

The wiring between terminals may be shown individually on simple diagrams, but with complicated systems such wiring is shown in the form of thick lines with the terminating ends entering and leaving just as if the wiring were arranged in looms. Clearly, Figures 1.5a and b are the wiring diagrams associated with the circuit shown in Figure 1.4. Although Figure 1.5a would be simple to wire without reference to the circuit diagram, Figure 1.5b would present a problem as it is shown if Figure 1.4 were not available.

In either case an alphanumeric (A1, GY56, f7, etc.) reference system is highly desirable, not only for ease of initial wiring, but also for fault location or the addition of circuitry at a later date. Both circuit and wiring diagrams should be cross-referenced with such a system (Figures 1.6a–c).

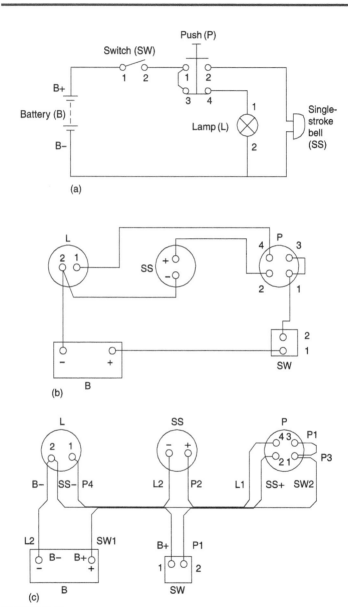

FIGURE 1.6   Schematic and wiring diagrams.

Note how, in Figure 1.6c, each termination is referenced with the destination of the conductor connected to it. Also note how much more easily a circuit diagram makes the interpretation of the circuits function.

CIRCUIT CONVENTION

It is probably sensible at this point to introduce the reader to circuit convention. This is simply a way of ensuring that circuit diagrams are more easily interpreted, and is achieved by drawing such diagrams in a de-energized state known as *normal*.

Hence, if we take a new motor starter out of its box, all of the coils, timers, overloads and contacts are said to be in their normal position. Figures 1.7a–d illustrate this convention as applied to relays and contactors.

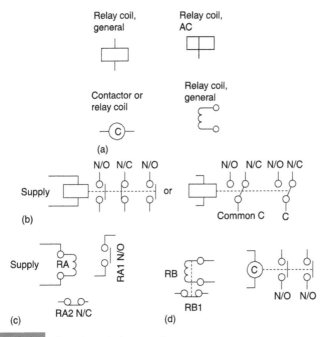

FIGURE 1.7   Contactor and relay conventions.

Note that, provided diagrams follow this accepted convention, it is unnecessary to label contacts normally open (N/O) or normally closed (N/C).

CONSTRUCTING AND INTERPRETING CIRCUIT DIAGRAMS

In order to construct or interpret a circuit/schematic diagram of the controls of a particular system, it is necessary to understand, in broad principles, how the system functions. A logical approach is needed, and it may take the novice some while before all 'clicks' into place.

Here is an example to consider.

Electronic valet

You work hard every day and return home late every evening. When you come in you look forward to a smooth scotch, a sit down and then a relaxing soak in a hot bath. If you were acquainted with electrical control systems you could arrange for these little luxuries to be automated as shown in Figure 1.8.

The system components are as follows:

TC	Typical 24 h time clock: TC1 is set to close at 2100 h.
KS	Key switch operated by front door key: momentary action, contacts open when key is removed.
T	Timer which can be set to close and open contacts T1 and T2 as required.
DD	Drinks dispenser with a sprung platform on which the glass sits. When energized, DD will dispense a drink into the glass. When the glass is removed, the platform springs up closing contacts 1 and 3 on DD1.
	DD1 Changeover contacts associated with DD.
	FS Normally closed float switch, which opens when the correct bath water level is reached.
	BFU (bath filling unit): electrically operated hot water valve.

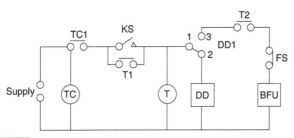

FIGURE 1.8   Electronic valet.

Let us now follow the system through:

1. At 9.00 pm or 2100 h the N/O contact TC1 on the time clock TC closes, giving supply to one side of the key switch and to the timer contact T1.

2. You arrive home and open the door with the key, which closes the N/O spring-return contacts on KS, thus energizing the timer T. The drinks dispenser DD is also energized via its own N/C contacts DD1 (1 and 2).

3. The timer T (now energized) instantly causes its own N/O contacts T1 to close, allowing supply to be maintained to T and DD via T1 (this is called a 'hold-on' circuit) when the key is removed from the key switch KS. N/O timer contacts T2 are set to close in say, 10 min. By the time you reach the lounge DD has poured your scotch.

4. When you remove the glass from the dispenser, DD1 contacts 1 and 2 open, and 1 and 3 close, de-energizing the dispenser and putting a supply to one side of the 10 min timed contacts T2.

5. You can now sit down, relax and enjoy your drink, knowing that shortly, contacts T2 will close and energize the bath filling unit BFU via the N/C float switch FS.

6. When the bath level is correct, the float switch FS opens and de-energizes BFU. You can now enjoy your bath.

7. One hour, say, after arriving home, the timer T will have completed its full cycle and reset, opening T1 and T2 and thus restoring the whole system to normal.

This system is, of course, very crude. It will work but needs some refinement. What if you arrive home early – surely you need not stay dirty and thirsty? How do you take a bath during the day without using the door key and having a drink? What about the bath water temperature? And so on. If you have already begun to think along these lines and can come up with simple solutions, then circuit/schematic diagrams should present no real problems to you.

Quiz controller

Here is another system to consider. Can you draw a circuit/schematic diagram for it? (A solution is given at the end of the book.)

The system function is as follows:

1. Three contestants take part in a quiz show. Each has a push-to-make button and an indicator lamp.
2. The quizmaster has a reset button that returns the system to normal.
3. When a contestant pushes his/her button, the corresponding lamp is lit and stays lit. The other contestants' lamps will not light.
4. The items of equipment are: a source of supply; a reset button (push-to-break); three push-to-make buttons; three relays each with 1 N/O and 2 N/C contacts and three signal lamps.

The resulting diagram is a good illustration of the use of an alpha-numeric system to show relay coils remote from their associated contacts.

HEATING AND VENTILATION SYSTEM

Figure 1.9 is part of a much larger schematic of the controls for the heating and ventilation system in a large hotel.

From the diagram it is relatively simple to trace the series of events that occur in this section of the system.

Clearly, there are four pumps: two boiler pumps and two variable temperature pumps. One of each of these pairs is a standby in the event of failure of the other; this will become clear as we interpret the scheme.

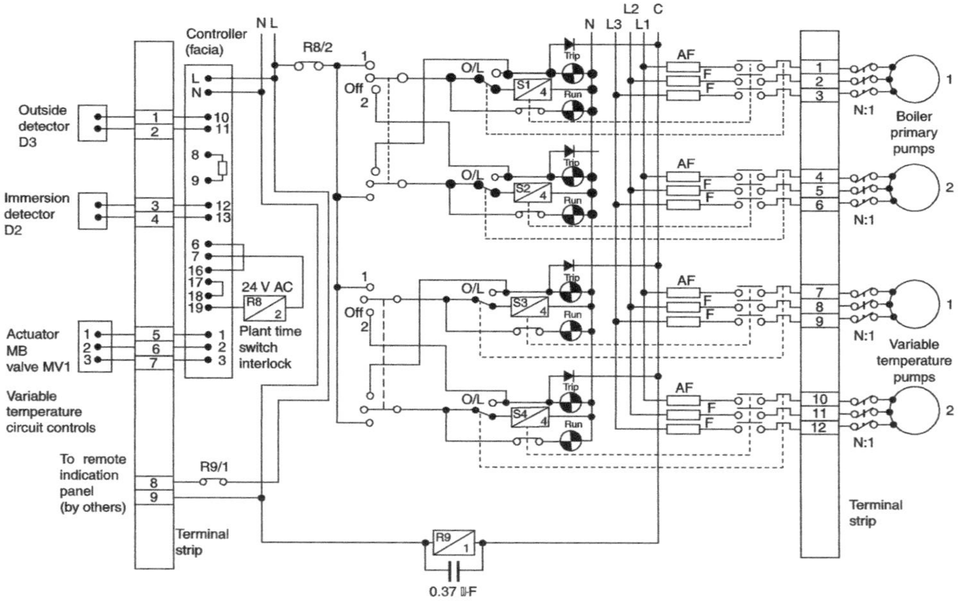

FIGURE 1.9 Heating and ventilation schematic diagram.

There is a controller (similar to the programmer of a central heating system) which receives inputs from two temperature sensors and operates an actuator valve and a time switch. There are two sets of linked, three-position switches and direct-on-line three-phase starters with single-phase coils S1/4, S2/4, S3/4 and S4/4 for the pumps. There is also run and trip indication for each pump.

Let us now follow the sequence of events:

1. The selector switches are set to, say, position 1.
2. The temperature sensors operate and the controller actuates valve MV1. If the 24 V time switch relay R8/2 is energized, then its N/O contacts R8/2 are closed, giving supply to the selector switches.
3. Starters S1/4 and S3/4 are energized via their respective overload (O/L) contacts; the main contacts close and the pumps start. Auxiliary contacts on the starters energize the run lamps.
4. If pump 1, say, were to overload, then the N/O O/L contacts would close, de-energizing S1/4 and shutting down pump 1, and supply would be transferred to starter S2/4 for pump 2 via the second linked switch. At the same time the trip lamp would come on and a supply via a diode and control cable C would be given to relay R9/1, operating its N/O contacts R9/1 to indicate a pump failure at a remote panel. The diode prevents back feeds to other trip lamps via the control cable C from other circuits.
5. The reader will see that the same sequence of events would take place if the selector switch were in position 2 in the first place.

RELAY LOGIC

In the last few pages we have investigated the use of relays for control purposes. Whilst this is perfectly acceptable for small applications, their use in more complex systems is now being superseded by programmable logic controllers (PLCs). However, before we discuss these in more detail, it is probably best to begin with a look at relay logic.

We have already discussed circuit convention with regard to N/O and N/C contacts, and in the world of logic these contacts are referred to as 'gates'.

And gates

If several N/O contacts are placed in series with, say, a lamp (Figure 1.10), it will be clear that contacts A *and* B *and* C must be closed in order for the lamp to light. These are known as **AND** gates.

Or gates

If we now rewire these contacts in parallel (Figure 1.11), they are converted to OR gates in that contact A *or* B *or* C will operate the lamp.

Combined gates

A combination of **AND** and **OR** systems is shown in Figure 1.12, and would be typical of, say, a remote start/stop control circuit for a motor. A *or* B *or* C will only operate the contactor coil if X *and* Y *and* Z are closed.

A simplification of any control system may be illustrated by a block diagram such as shown in Figure 1.13, where the input may be achieved by

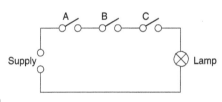

FIGURE 1.10 AND gates.

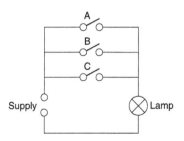

FIGURE 1.11 OR gates.

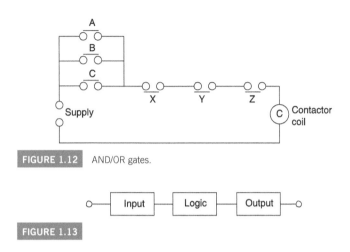

FIGURE 1.12 AND/OR gates.

FIGURE 1.13

the operation of a switch or sensor, the logic by relays, coils, timers, etc., and the outputs in the form of lamps, heaters, sounders, contactors, etc.

PROGRAMMABLE LOGIC CONTROLLERS

With complex control requirements, the use of electro-mechanical relays is somewhat cumbersome, and most modern systems employ PLCs. In basic terms these do no more than relays (i.e. they process the input information and activate a corresponding output). Their great advantage, however, is in the use of microelectronics to achieve the same end. The saving in space and low failure rate (there are no moving parts) make them very desirable. A typical unit for, say, 20 inputs (I) and 20 outputs (O), referred to as a 40 I/O unit, would measure approximately 300 mm by 100 mm by 100 mm, and would also incorporate counters, timers, internal coils, etc.

A PLC is programmed to function in a specified way by the use of a keyboard and a display screen. The information may be programmed directly into the PLC, or a chip known as an EPROM may be programmed remotely and then plugged into the PLC. The programming method uses 'ladder logic'. This employs certain symbols, examples of which are shown in Figure 1.14. These symbols appear on the screen as the ladder diagram is built up.

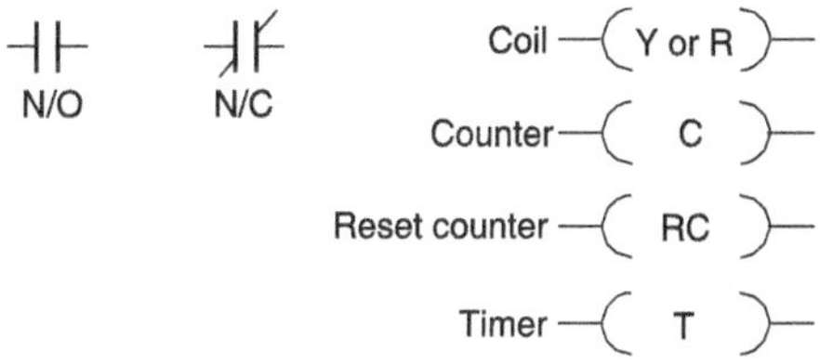

FIGURE 1.14   Ladder logic.

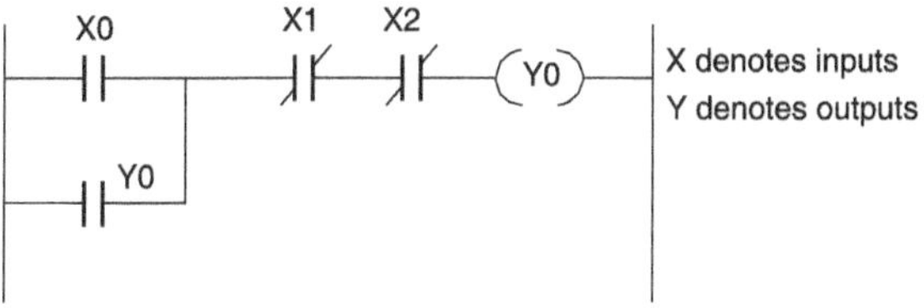

FIGURE 1.15

Here are some examples of the use of ladder logic.

Motor control

Figure 1.15 illustrates a ladder logic diagram for a motor control circuit (no PLC involved here). Closing the N/O contacts X0 gives supply to the motor contactor coil Y0 via N/C stop buttons X1 and X2. Y0 is held on via its own N/O contact Y0 when X0 is released. The motor is stopped by releasing either X1 or X2.

Packing control

Figure 1.16 shows the basic parts of a packing process. An issuing machine ejects rubber balls into a delivery tube and thence into boxes on a turntable. A photoswitch senses each ball as it passes. Each box holds 10 balls and the turntable carries 10 boxes.

Clearly, the issuing machine must be halted after the 10th ball, and time allowed for all balls to reach their box before the turntable revolves to bring another box into place. When the 10th box has been

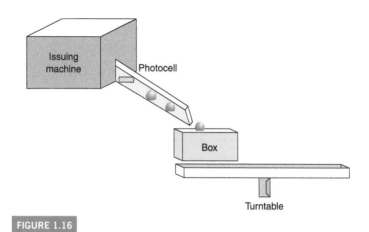

Issuing
machine

Photocell

Box

Turntable

FIGURE 1.16

filled, the system must halt and a warning light must be energized to indicate that the process for that batch is completed. When new boxes are in place the system is restarted by operating an N/C manual reset button.

This system is ideal for control by a PLC with its integral counters and timers. Figure 1.17 shows an example of the ladder logic for this system using the following:

X0	N/O photocell switch: closes as ball passes.
X1	N/C manual reset button.
Y0	Output supply to issuing machine.
Y1	Output supply to turntable.
Y2	Output supply to warning light.
C0	Internal counter set to 10 with one N/C and two N/O contacts.
C1	Internal counter set to 10 with one N/C and one N/O contacts.
T0	Timer set for 5 s with one N/O contact.
T1	Timer set for 5 s with one N/C contact.
RC	Reset counter: resets counter when supply to it is cut.

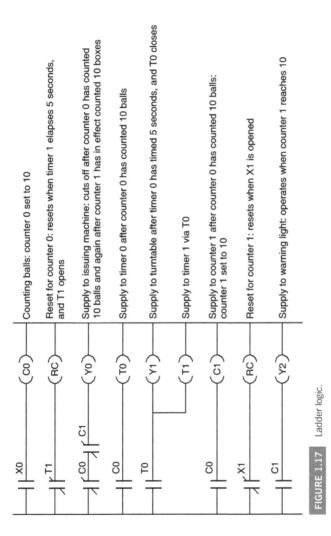

Ladder rung	Description
X0 — (C0)	Counting balls: counter 0 set to 10
T1 — (RC)	Reset for counter 0: resets when timer 1 elapses 5 seconds, and T1 opens
C0 C1 — (Y0)	Supply to issuing machine: cuts off after counter 0 has counted 10 balls and again after counter 1 has in effect counted 10 boxes
C0 — (T0)	Supply to timer 0 after counter 0 has counted 10 balls
T0 — (Y1)	Supply to turntable after timer 0 has timed 5 seconds, and T0 closes
— (T1)	Supply to timer 1 via T0
C0 — (C1)	Supply to counter 1 after counter 0 has counted 10 balls: counter 1 set to 10
X1 — (RC)	Reset for counter 1: resets when X1 is opened
C1 — (Y2)	Supply to warning light: operates when counter 1 reaches 10

FIGURE 1.17 Ladder logic.

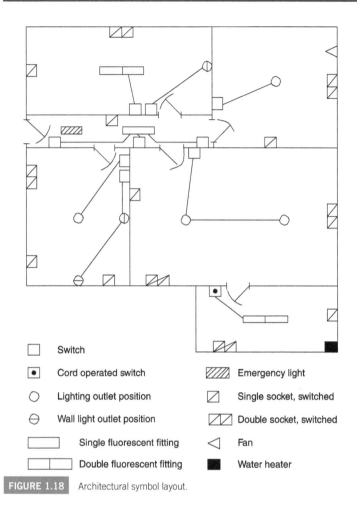

☐	Switch		
▣	Cord operated switch	▨	Emergency light
○	Lighting outlet position	◺	Single socket, switched
⊖	Wall light outlet position	◹◺	Double socket, switched
▭	Single fluorescent fitting	◁	Fan
▭▭	Double fluorescent fitting	■	Water heater

FIGURE 1.18 Architectural symbol layout.

Fault location

Another major advantage of the use of PLCs for controlling systems is the relative ease of fault location. In the event of system failure, the keyboard and screen unit is plugged into the PLC and the condition of the system is displayed in ladder logic on the screen. Then, for example, any contact that is in the wrong position will show up.

DRAWING EXERCISES

1. Using BS EN 60617 architectural symbols, draw *block* diagrams of the following circuits:

   (a) A lighting circuit controlled by one switch, protected by a fuse, and comprising three tungsten filament lamp points, two double fluorescent luminaires, and one single fluorescent luminaire.

   (b) A lighting circuit controlled by two-way switches, protected by a fuse, and comprising three floodlights.

   (c) A lighting circuit controlled by two-way switches, and one intermediate switch, protected by a circuit breaker, and comprising three spotlights. One of the two-way switches is to be cable operated.

   (d) A ring final circuit protected by a circuit breaker, and comprising six double switched socket outlets and two single switched socket outlets.

2. Replace the symbols shown in Figure 1.18 with the correct BS EN 60617 symbols.

Solutions are given at the end of the book.

CHAPTER 2

Wiring Systems

Important terms/topics covered in this chapter:

- BS 7671 conductor identification
- Ring, radial and distribution circuits and systems
- Emergency lighting, alarm and security systems
- Call systems
- Motor starting systems
- Central heating systems
- Extra low voltage systems
- Domestic telephone systems

By the end of this chapter the reader should:

- be aware of the various wiring systems, i.e. ring, radial and distribution,
- understand the need for installation diagrams to have a circuit iden-tification system,
- understand 'open' and 'closed' wiring systems,
- know that knowledge of how a system functions helps the interpre-tation of diagrams.

In this chapter we will investigate a selection of the many wiring sys-tems employed in modern installations. Some of these systems are simple to understand and require little explanation. Others of a more complex nature should now, in the light of the reader's new-found knowledge of diagrams, etc., present only minor problems of inter-pretation.

It should be noted that diagrams for LV systems rarely indicate conductor colours, it is more likely that control circuit and ELV system diagrams will show these. Table 2.1 shows the colours/alphanumeric references required by the IET Wiring Regulations.

RADIAL SYSTEMS

Any system which starts from the supply point and either radiates out like the spokes of a wheel, or extends from one point to another in the form of

Table 2.1   Colours and Alphanumeric References

	Conductor	Letter/Number	Colour
Single-phase AC			
	Line	L	Brown
	Neutral	N	Blue
Three-phase AC			
	Line 1	L1	Brown
	Line 2	L2	Black
	Line 3	L3	Grey
	Neutral	N	Blue
Control wiring or ELV			
	Line	L	Brown, Black, Red, Orange, Yellow, Violet, Grey, White, Pink or Turquoise
	Neutral		Blue
For all systems			
	Protective		Green-yellow

a chain, is a radial system. Figures 2.1, 2.2 and 2.3 illustrate such systems as applied to lighting and power circuits.

It should be noted that BS EN 60617 architectural symbols are not often shown in this fashion; it is usual to see them used in conjunction with building plans. This will be discussed later.

RING CIRCUITS

These circuits start at the supply point, loop from point to point and return to the same terminals they started from. They are most popular in domestic premises, where they are referred to as ring final circuits.

However, such systems are also used in factories where overhead busbar trunking is in the form of a ring, or for supply authority networks (Figures 2.4 and 2.5).

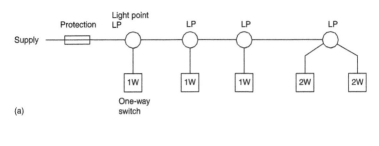

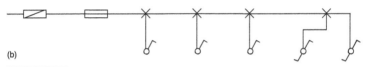

FIGURE 2.1   Radial lighting circuit using (a) representative, (b) architectural symbols.

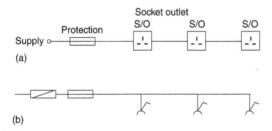

FIGURE 2.2   Radial socket outlet circuit using (a) representative, (b) architectural symbols.

DISTRIBUTION SYSTEMS

Such systems are many and varied, but they are quite simple to understand as they tend to follow the ring and radial concepts.

Take, for example, the UK electricity system. Regardless of who owns this or that part of it, the system functions in the following stages: generation, transmission and distribution. Generated electricity is transmitted over vast distances around the United Kingdom in a combination of ring and radial circuits to points of utilization, where it is purchased by the distribution network operators (DNOs) and distributed to their customers. Once again these systems are in ring or radial forms.

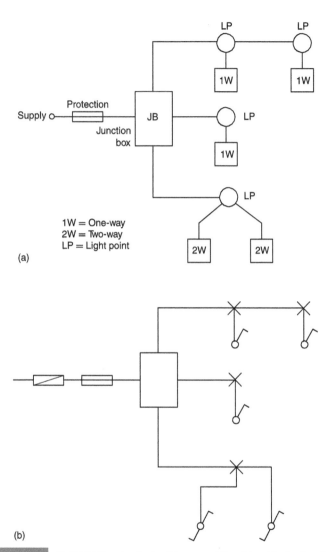

FIGURE 2.3 Radial lighting circuit using (a) representative, (b) architectural symbols.

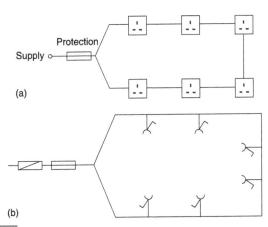

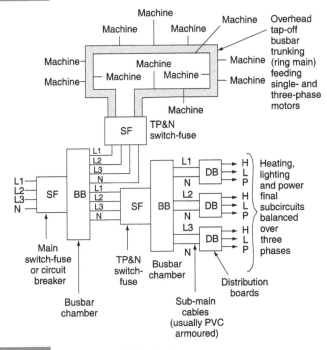

FIGURE 2.4 Ring final circuit using (a) representative, (b) architectural symbols.

FIGURE 2.5 Layout of industrial installation.

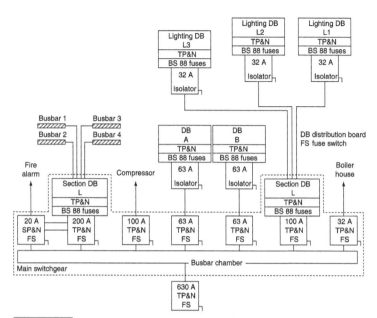

FIGURE 2.6 Distribution system, block type.

Probably more familiar to the installation electrician is the distribution system in an industrial or commercial environment. Here one finds radial circuits originating from the intake position and feeding distribution boards (DBs), from which are fed either more DBs or final circuits. Diagrams for such systems may be of the block type (Figure 2.6) or of the interconnection type (Figure 2.7).

Note how much more detail there is on the section of the drawing shown in Figure 2.7. Cable sizes and types are shown, together with cable lengths (23, 26 m, etc.). Details at each DB indicate current loading (CC), approximate maximum demand (AMD), voltage drop (VD), earth loop impedance (ELI) and prospective fault current (PFC).

With the larger types of installation, an alphanumeric system is very useful for cross-reference between block diagrams and floor plans showing architectural symbols. Figure 2.8 (see page 30) shows such a system.

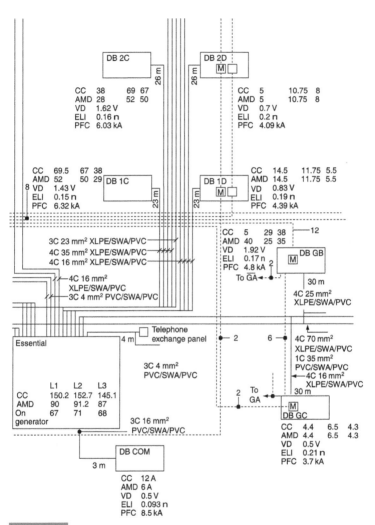

FIGURE 2.7 Distribution system, interconnection type.

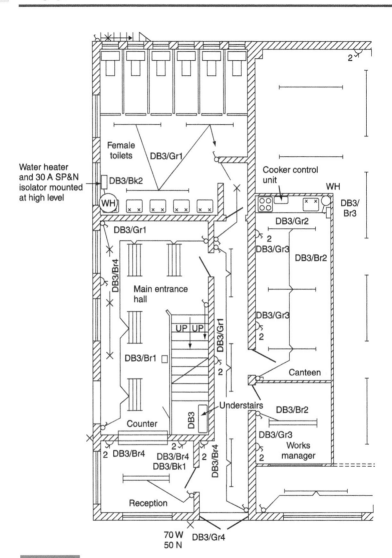

FIGURE 2.8 Example floor plan.

Distribution board 3 (DB3) under the stairs would have appeared on a diagram such as Figure 2.7, with its final circuits indicated. The floor plan shows which circuits are fed from DB3, and the number and phase colour of the protection. For example, the fluorescent lighting in the main entrance hall is fed from protective device number 1 on the grey phase of DB3/Gr1, and is therefore marked DB3/Br1. Similarly, the water heater circuit in the female toilets is fed from protective device number 2 on the black phase (i.e. DB3/Bk2).

Figures 2.9, 2.10 and 2.11 illustrate a simple but complete scheme for a small garage/workshop. Figure 2.9 is an isometric drawing of the garage and installation, from which direct measurements for materials may be taken. Figure 2.10 (see page 33) is the associated floor plan, which cross references with the DB schedule and interconnection details shown in Figure 2.11 (see page 34).

A similar diagram to Figure 2.11 is shown for part of the ventilation system in a commercial premises in Figure 2.12 (see page 35).

EMERGENCY LIGHTING SYSTEMS

These fall into two categories: maintained and non-maintained. Both these systems may be utilized by individual units or by a centralized source.

Maintained system

In this system the emergency lighting unit is energized continuously via a step-down transformer, and in the event of a mains failure it remains illuminated via a battery (Figure 2.13, see page 36).

Non-maintained system

Here the lighting units remain de-energized until a mains failure occurs, at which time they are illuminated by a battery supply (Figure 2.14).

It should be noted that modern systems use electronic means to provide the change-over from mains to battery supply. The contactor method, however, serves to illustrate the principle of operation.

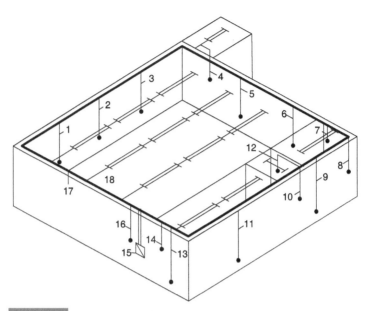

FIGURE 2.9 Isometric drawing for garage/workshop.

1 – Three-phase supply to ramp: 20 mm conduit

2 – Single-phase supply to double sockets: 20 mm conduit. Also 3, 5, 6, 9, 11, 13

4 – Single-phase supply to light switch in store: 20 mm conduit

7 – Single-phase supply to light switch in compressor: 20 mm conduit

8 – Three-phase supply to compressor: 20 mm conduit

10 – Single-phase supply to heater in WC: 20 mm conduit

12 – Single-phase supply to light switch in WC: 20 mm conduit

14 – Single-phase supply to light switch in office: 20 mm conduit

15 – Main intake position

16 – Single-phase supplies to switches for workshop lights: 20 mm conduit

17 – 50 mm × 50 mm steel trunking

18 – Supplies to fluorescent fittings: 20 mm conduit.

SECURITY AND FIRE ALARM SYSTEMS

As with emergency lighting, modern security and fire alarm systems are electronically controlled, and it is not the brief of this book to investigate electronic circuitry. However, as with the previous section, the basic principle of operation can be shown by electro-mechanical means.

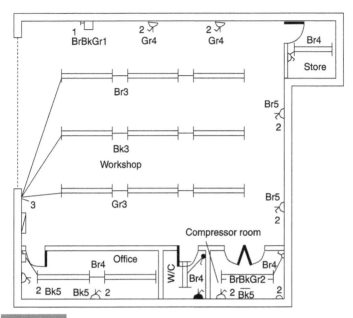

FIGURE 2.10   Floor plan for garage/workshop.

Both security and fire alarm systems are basically the same in that various sensors are wired to a control panel, which in turn will activate an alarm in the event of sensor detection (Figures 2.15 and 2.16). Some modern panels have the facility for incorporating both systems in the same enclosure.

The principles of operation are as follows.

Open-circuit system

In this system the call points (sensors, detectors, etc.) are wired in parallel such that the operation of any one will give supply to the relay RA and the sounder via the reset button (Figure 2.17). N/O contacts RA1 will then close, holding on the relay and keeping to the sounder via these contacts. This hold-on facility is most important as it ensures that the sounder is not interrupted if any attempt is made to return the activated call point to its original off position.

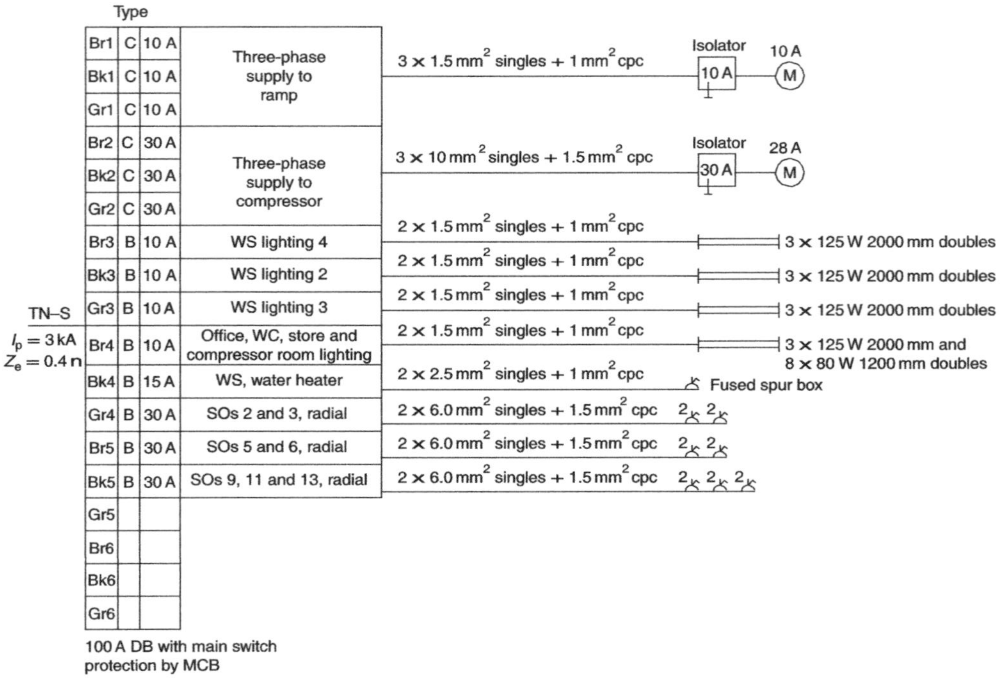

Type

				Cable	Load
Br1	C	10 A	Three-phase supply to ramp	3 × 1.5 mm² singles + 1 mm² cpc	Isolator 10 A → 10 A (M)
Bk1	C	10 A			
Gr1	C	10 A			
Br2	C	30 A	Three-phase supply to compressor	3 × 10 mm² singles + 1.5 mm² cpc	Isolator 28 A → 30 A (M)
Bk2	C	30 A			
Gr2	C	30 A			
Br3	B	10 A	WS lighting 4	2 × 1.5 mm² singles + 1 mm² cpc	3 × 125 W 2000 mm doubles
Bk3	B	10 A	WS lighting 2	2 × 1.5 mm² singles + 1 mm² cpc	3 × 125 W 2000 mm doubles
Gr3	B	10 A	WS lighting 3	2 × 1.5 mm² singles + 1 mm² cpc	3 × 125 W 2000 mm doubles
Br4	B	10 A	Office, WC, store and compressor room lighting	2 × 1.5 mm² singles + 1 mm² cpc	3 × 125 W 2000 mm and 8 × 80 W 1200 mm doubles
Bk4	B	15 A	WS, water heater	2 × 2.5 mm² singles + 1 mm² cpc	Fused spur box
Gr4	B	30 A	SOs 2 and 3, radial	2 × 6.0 mm² singles + 1.5 mm² cpc	2 2
Br5	B	30 A	SOs 5 and 6, radial	2 × 6.0 mm² singles + 1.5 mm² cpc	2 2
Bk5	B	30 A	SOs 9, 11 and 13, radial	2 × 6.0 mm² singles + 1.5 mm² cpc	2 2 2
Gr5					
Br6					
Bk6					
Gr6					

TN–S
$I_p = 3$ kA
$Z_e = 0.4$ n

100 A DB with main switch protection by MCB

FIGURE 2.11 Details of connection diagram for garage/workshop.

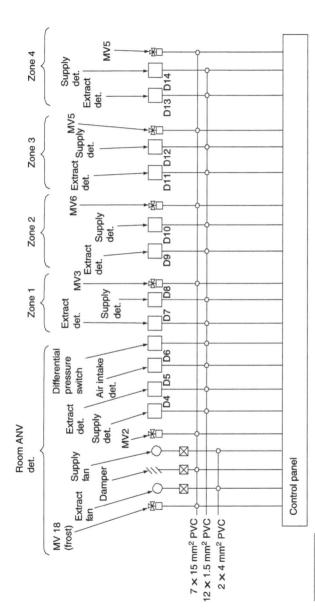

FIGURE 2.12 Connection diagram for part of ventilation scheme (det.: detector).

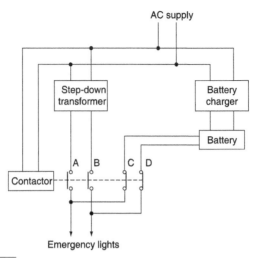

FIGURE 2.13 Maintained system.

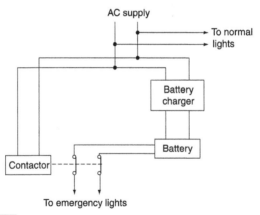

FIGURE 2.14 Non-maintained system.

Fire alarm systems are usually wired on an open circuit basis, with a two-wire system looped from one detector to the next, terminating across an end-of-line resistor (EOLR). This provides a circuit cable monitoring facility; the EOLR is of sufficiently high value to prevent operation of the alarm. Figure 2.18 shows a typical connection diagram.

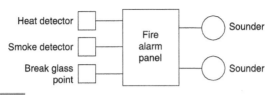

FIGURE 2.15   Block diagram for fire alarm system.

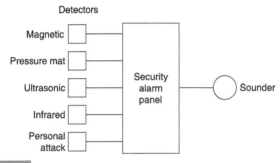

FIGURE 2.16   Block diagram for security alarm system.

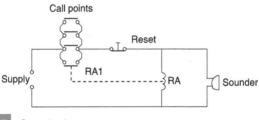

FIGURE 2.17   Open circuit.

Closed-circuit system

This system has the call points wired in series, and the operation of the reset button energizes the relay RA (Figure 2.19). Normally open (N/O) contacts RA1 close and normally closed (N/C) contacts RA2 open, the relay remaining energized via contacts RA1 when the reset button is released. The alarm sounder mute switch is then closed, and the whole system is now set up.

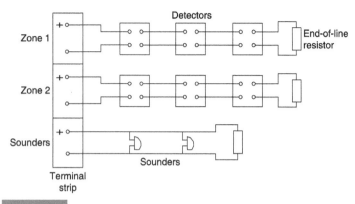

FIGURE 2.18 Fire alarm system.

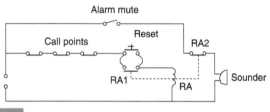

FIGURE 2.19 Closed circuit.

An interruption of the supply to relay RA, by operation of any call point, will de-energize the relay, open RA1 and close RA2, thus actuating the alarm sounder. The system can only be cancelled and reset by use of the reset button.

The closed system is quite popular, as it is self-monitoring in that any malfunction of the relay or break in the call point wiring will cause operation of the system as if a call point had been activated.

Intruder alarm systems tend, in the main, to be based on the closed-circuit type. Figure 2.20 shows the connection diagram for a simple two-zone system with tamper loop and personal attack button.

The tamper loop is simply a continuous conductor wired to a terminal in each detector in the system. It is continuously monitored irrespective

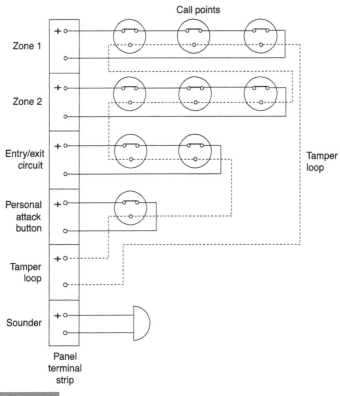

FIGURE 2.20   Security alarm system.

of whether the alarm system is switched on or off, and if interrupted will cause immediate operation of the alarm.

The entry/exit circuit is usually confined to the front and/or back doors. The facility exists to alter the time delay between setting the system and exiting, and between entering and switching the system off. This adjustment is made inside the control panel.

All security and fire alarm systems should have battery back-up with charging facilities.

Both fire and emergency lighting systems are Safety Services and are dealt with, generally, in Chapter 56 of the IET Wiring Regulations. More detailed requirements are to be found in BS 5266 for emergency lighting and BS 5839 for fire protection.

CALL SYSTEMS

Once again these fall into different categories, such as telephone systems and page and bleeper systems. However, the typical nurse-call variety, which uses push-buttons and lamp indication, is probably the most popular.

With this type, each room is equipped with a call button of some description, a patient's reassurance light and a cancel button. Outside each room is an indicator light, and at strategic points in the building are zone sounders. Centrally located is a display panel which incorporates a sounder and an indication of which room is calling.

Figure 2.21 illustrates, in a simple form, the principle of operation of such a system. This system should, by now, be quite familiar to the reader; it is simply another variation of the hold-on circuit. Any patient pushing a call button energizes the corresponding relay in the main control panel, which is held on by a pair of N/O contacts. At the same time the reassurance, room and panel lights 1, 2 and 3 are all illuminated. The zone and panel sounders are energized via the relay's other pair of N/O contacts.

It is usual to locate cancel buttons in patient's rooms only, as this ensures that staff visit the patient in question.

All of the systems just dealt with are generally supplied from an extra low-voltage (ELV) source and are known as Band I circuits. These should not be contained within the same wiring system as low voltage (LV) systems which are Band II circuits unless they are insulated to the highest voltage present or, if in trunking, segregated by a partition.

MOTOR STARTER CIRCUITS

No book on wiring systems would be complete without reference to control circuits for motor starters. Here we will look at direct-on-line

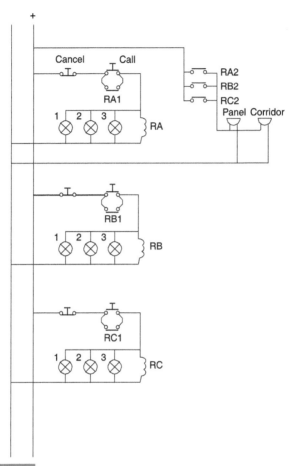

FIGURE 2.21 Nurse-call system.

(DOL) and star-delta starters. Once more, the good old hold-on circuit is employed in both types.

Direct-on-line starter

Both single- and three-phase types use the same circuit, as illustrated in Figures 2.22 and 2.23. It is important to note that if a 230 V coil were

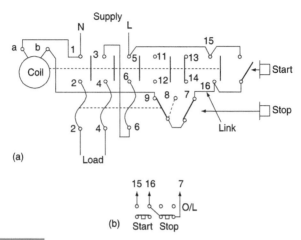

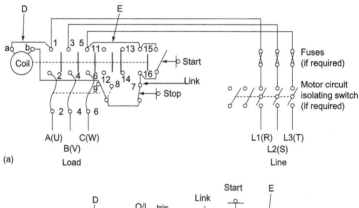

FIGURE 2.22 (a) Direct-on-line starter, single-phase. (b) Connections for remote push-button (start/stop) control: omit link and connect as shown.

FIGURE 2.23 (a) Direct-on-line starter, three-phase. (b) Schematic diagram. Control circuit supply: for line to line, connect as shown; for line to neutral, omit connection D and connect neutral to terminal a; for separate supply, omit D and E, and connect separate coil supply to terminals a and 15.

to be used instead of a 400 V type, the coil connections would require a neutral conductor in the starter.

Star-delta starter

Figures 2.24 and 2.25 show the wiring and schematic diagrams for a star-delta starter. This is clearly a more complicated system than the DOL type. However, the control system is essentially the same. On start-up, the star contactor, electronic timer and the main contactor are all energized. At the end of the set time the supply to the star contactor is removed and given to the delta contactor, the main contactor remaining energized the whole time. Reference to Figure 2.25 will indicate how the system functions.

1. When the start is pushed, supply is given to the star contactor Ⓐ and the timer ET, from L1 to L3, and hence all contacts marked λ associated with the star contactor will change position. The supply to Ⓐ and ET is maintained, after the start is released, via star contacts 11 and 12, ET contacts OR-Y, and star contacts 15 and 16. The main contactor Ⓜ is also energized via star contacts 11 and 12, and M15, and thereafter maintained via its own contacts M15 and 16.
2. The motor has of course started, and after a predetermined time delay ET operates and its contacts OR-Y change to OR-L. This cuts off supply to the star contactor Ⓐ and also ET. All star contacts return to normal, and ET resets OR-11 to OR-Y.
3. Supply is now given to the delta contactor Ⓐ via M15 and 16, OR-Y, and star contacts λ 13 and 14. Delta contacts Δ 13 and 14 open and prevent further energization of the star contactor.

The reader will notice that the line and load terminal markings in Figure 2.24a show letters in brackets; these are the European equivalents.

CENTRAL HEATING SYSTEMS

It would be impossible in such a small book to deal with the vast number of modern systems and variations that are currently available. We will therefore take a look at the two most basic arrangements: the pumped central heating (CH) and gravity-fed hot water (HW) system, and the

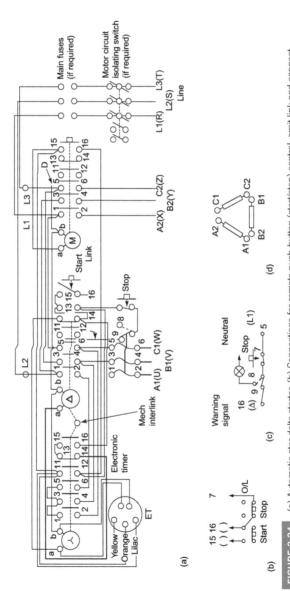

FIGURE 2.24 (a) Automatic star-delta starter. (b) Connections for remote push-button (start/stop) control: omit link and connect as shown. (c) Connections for trip warning. (d) Motor windings: connect to appropriate terminals on starter.

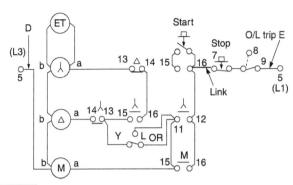

FIGURE 2.25 Schematic diagram. Control circuit supply: for line to line, connect as shown; for line to neutral, omit connection D and connect neutral to terminal b; for separate supply, omit D and E, and connect separate coil supply to terminals b and 9 connections for remote pilot switch control: remove connection 14 to 15 on delta contactor; connect between 14 and 16 on M contactor to terminal 14 on delta contactor; connect pilot switch in place of connection E.

fully pumped system with mid-position valve. It must be remembered that, whatever the system, it is imperative that the wiring installer has a knowledge of the function of the system in order for him/her to do a competent job.

Pumped CH and gravity HW

This system comprises a boiler with its own thermostat to regulate the water temperature, a pump, a hot water storage tank, a room thermostat and some form of timed programmer. The water for the HW (i.e. the taps, etc.) is separate from the CH water, but the boiler heats both systems.

Figure 2.26 shows such a system. From the diagram it might appear that when the requirement for HW is switched off at the programmer, the CH cannot be called for as the boiler has lost its feed. In fact, such programmers have a mechanical linkage between the switches: HW is allowed without CH, but selection of CH automatically selects HW also.

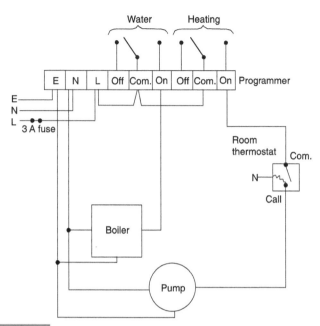

FIGURE 2.26   Gravity primary and pumped heating.

Note the little heating element in the room thermostat; this is known as an accelerator. Its purpose is to increase the sensitivity of the thermostat; manufacturers claim that it increases the accuracy of the unit to within one degree Celsius. The inclusion of an accelerator (if required) does mean an extra conductor for connection to neutral.

Fully pumped system

Two additional items are required for this system: a cylinder thermostat and a mid-position valve. In this system HW and CH can be selected independently. The mid-position valve has three ports: a motor will drive the valve to either HW only, CH only or HW and CH combined. With this system the boiler and pump always work together. Figure 2.27 illustrates the system, and Figure 2.28 (see page 48) shows the internal connections of a mid-position valve.

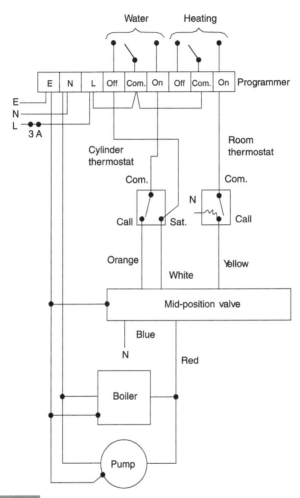

FIGURE 2.27 Fully pumped system.

Some difficulties may be experienced in wiring when the component parts of the system are produced by different manufacturers. In this case it is probably best to draw one's own wiring diagram from the various details available.

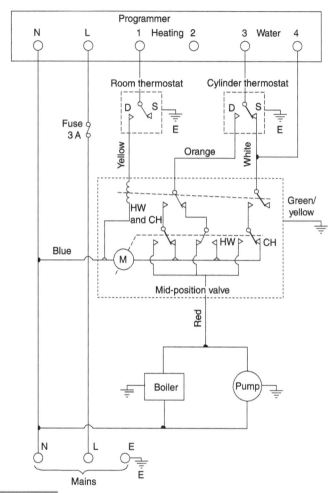

FIGURE 2.28 | Internal connections of mid-position valve.

EXTRA LOW-VOLTAGE LIGHTING

These systems, incorrectly referred to as low-voltage lighting (low voltage is 50–1000 V AC), operate at 12 V AC. They employ tungsten halogen dichroic lamps, which have a very high performance in comparison with

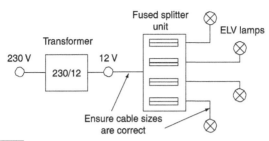

FIGURE 2.29 ELV lighting system block diagram.

240 V halogen lamps. For example, a 50 W dichroic lamp has the same intensity as a 150 W PAR lamp.

Extra low-voltage (ELV) lighting is becoming very popular, especially for display purposes. There is very little heat emission, the colour rendering is excellent and energy consumption is very low.

The 12 V AC to supply the lamps is derived from a 230 V/12 V transformer specially designed to cater for the high starting surges, and only these types should be used. The voltage at each lamp is critical: 0.7 V overvoltage can cause premature ageing of the lamp, and 0.7 V undervoltage will reduce the light output by 30%. Hence the variation in voltage must be avoided.

To achieve this, leads and cables must be kept as short as possible, and the correct size used to avoid excessive voltage drop. When several lamps are to be run from one transformer, it is advisable to use a fused splitter unit rather than to wire them in a parallel chain (Figure 2.29).

It is important to remember that, for example, a 50 W ELV lamp will draw 4.17 A from the 12 V secondary of the transformer (I = P/V). Although a 1.0 mm^2 cable will carry the current, the voltage drop for only 3 m of this cable will be 0.55 V.

DOMESTIC TELEPHONE SYSTEMS

Extensions to domestic telephone systems are extremely easy as each extension socket is wired in parallel to the one previous (Figure 2.30).

Master   Secondary sockets

FIGURE 2.30 Telephone system block diagram.

The master socket is the first socket in any installation and contains components to allow telephones to

- be removed without causing problems at the exchange
- stop surges on the line such as lightning strikes
- prevent the telephone making partial ringing noises when dialling.

Connection to the master socket is not permitted, except by use of an adaptor plug and extension cable.

Extension or secondary sockets house only terminals.

Secondary sockets

The number of these is unlimited but the number of modern telephones or ringing devices (e.g. extension bells) connected at any one time is limited to four. More than this and telephones may not ring or even work.

Cable

The cable used should comply with BT specification CW1308, which is 1/0.5 mm and ranges from four-core (two pairs) to 40-core (20 pairs). It is not usual for secondary sockets to require any more than four cores.

Wiring

Wiring may be flush or surface but kept clear of the low-voltage electrical system by at least 50 mm. No more than 100 m of cable should be used overall and the length between the master socket and the first extension socket should not be more than 50 m.

Connection to the modern insulation displacement connectors (IDCs) is made using a special tool provided with each socket. The connection requirements are as shown in Figure 2.31.

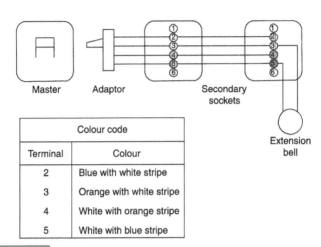

Colour code	
Terminal	Colour
2	Blue with white stripe
3	Orange with white stripe
4	White with orange stripe
5	White with blue stripe

FIGURE 2.31 Secondary socket wiring.

CHAPTER 3

Testing and Test Instruments

Important terms/topics covered in this chapter:

- Inspection
- Electrical quantities
- Selection of test instruments
- Approved test lamps and voltage indicators
- Care of instruments
- Tests and testing

At the end of this chapter the reader should:

- know the instrument ranges and measurement quantities,
- be able to identify the correct certification documents to be completed,
- be aware of the importance of having accurate instrumentation,
- know the correct method of proving a circuit is dead and safe to work on,
- be able to list the relevant tests and the sequence in which they should be conducted,
- understand the theory behind, and the methods of, testing,
- understand and interpret test results.

Whilst this section deals with testing, it should be remembered that such action should always be preceded by inspection as this may reveal faults, damage, deterioration, etc., that cannot be detected by testing alone.

MEASUREMENT OF ELECTRICAL QUANTITIES

As the reader will know, the basic electrical quantities which need to be measured in the world of the installation electrician are current, voltage and resistance. The units of these quantities are the ampere, the volt and the ohm, respectively.

Paradoxically, however, the range and complexity of the instruments available to measure these three fundamental quantities are enormous.

18th Edition IET Wiring Regulations: Wiring Systems and Fault Finding for Installation Electricians. 978-1-138-60611-1.
© Brian Scaddan. Published by Taylor & Francis. All rights reserved.

Table 3.1		
Test	**Range**	**Type of Instrument**
1. Continuity of ring final conductors	0.05–0.8 Ω	Low reading ohmmeter
2. Continuity of protective conductors	2–0.005 Ω or less	Low reading ohmmeter
3. Earth electrode resistance	Any value over about 3 or 4 Ω	Special ohmmeter
4. Insulation resistance	Infinity to less than 1 MΩ	High reading ohmmeter
5. Polarity	None	Ohmmeter, bell, etc.
6. Earth fault loop impedance	0–2000 Ω	Special ohmmeter
7. Operation of RCD	5–500 mA	Special instrument
8. Prospective fault current	2 A–20 kA	Special instrument

So where does one start in order to make a choice of the most suitable types?

Let us look at the range of quantities that an electrician is likely to encounter in the normal practice of his/her profession. If we take the sequence of the more commonly used tests prescribed by the IET Wiring Regulations, and assign typical values to them, we can at least provide a basis for the choice of the most suitable instruments. It will be seen from Table 3.1 that all that is required is an ohmmeter of one sort or another, a residual current device (RCD) tester and an instrument for measuring prospective fault current (PFC).

SELECTION OF TEST INSTRUMENTS

It is clearly most sensible to purchase instruments from one of the established manufacturers rather than to attempt to save money by buying cheaper, lesser known brands. Also, as the instruments used in the world of installation are bound to be subjected to harsh treatment, a robust construction is all important. Reference should be made to the HSE Guidance Note GS38 'Electrical test equipment for use by electricians'.

Many of the well-known instrument companies provide a dual facility in one instrument, for example PFC and loop impedance, or insulation resistance and continuity. Hence it is likely that only one, three or four instruments would be needed, together with an approved test lamp (a subject to be dealt with in the next section).

Now, which type to choose, analogue or digital? There are merits in both varieties, and the choice should not be determined just by expense or a reluctance to use 'new-fangled electronic gadgetry'! This attitude has, however, become a thing of the past, and an analogue variety would probably only be purchased second hand! Accuracy, ease of use and robustness, together with personal preference, are the all-important considerations.

APPROVED TEST LAMPS AND VOLTAGE INDICATORS

Search your tool boxes; find, with little difficulty one would suspect, your 'neon screwdriver' or 'testascope'; locate a very deep pond; and drop it in!

Imagine actually allowing electric current at low voltage (50–1000 V AC) to pass through one's body in order to activate a test lamp! It only takes around 10–15 mA to cause severe electric shock, and 50 mA (1/20th of an ampere) to kill.

Apart from the fact that such a device will register any voltage from about 5 V upwards, the safety of the user depends entirely on the integrity of the current-limiting resistor in the unit. An electrician received a considerable shock when using such an instrument after his apprentice had dropped it in a sink of water, simply wiped it dry and replaced it in the tool box. The water had seeped into the device and shorted out the resistor.

It should be said, however, that some modern types are available that are proximity devices and can give an indication of the presence of voltage. Such a device should **never** be relied on to suggest that a circuit is dead and safe to work on. It could be useful to indicate if a fuse had operated by placing it either side.

An approved test lamp should be of similar construction to that shown in Figure 3.1.

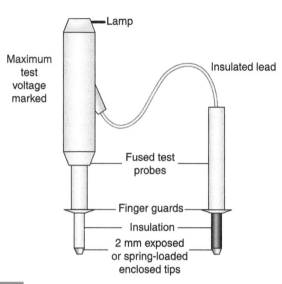

Lamp

Maximum test voltage marked

Insulated lead

Fused test probes

Finger guards

Insulation

2 mm exposed or spring-loaded enclosed tips

FIGURE 3.1 Approved test lamp.

ACCIDENTAL RCD OPERATION

It has long been the practice when using a test lamp to probe between line and earth for indication of a live supply on the line terminal. However, this can now prevent a problem where RCDs exist in the circuit, as of course the test is applying a deliberate line to earth fault.

Some test lamps have LED indicators, and the internal circuitry of such test lamps limits the current to earth to a level below that at which RCD will operate. The same limiting effect applies to multimeters. However, it is always best to check that the testing device will have no effect on RCDs.

CALIBRATION, ZEROING AND CARE OF INSTRUMENTS

Precise calibration of instruments is usually well outside the province of the electrician, and would normally be carried out by the manufacturer or a local service representative. A check, however, can be made by the user to determine whether calibration is necessary by comparing readings with an instrument known to be accurate, or by measurement of known values of voltage, resistance, etc., available on units now on the market and advertised as 'check boxes'.

It may be the case that readings are incorrect simply because the instrument is not zeroed or nulled before use, or because the internal battery needs replacing. Most modern instruments have battery condition indication, and of course this should never be ignored.

Always adjust any selection switches to the off position after testing. Too many instrument fuses are blown when, for example, a multimeter is inadvertently left on the ohms range and then used to check for mains voltage.

The following procedure may seem rather basic but should ensure trouble-free testing:

1. Check test leads for obvious defects.
2. Zero/null the instrument (where required).
3. Select the correct range for the values anticipated. If in doubt, choose the highest range and gradually drop down.
4. Make a record of test results, if necessary.
5. Return switches/selectors to the off position.
6. Replace instrument leads in carrying case.

CONTINUITY OF PROTECTIVE CONDUCTORS

All protective conductors, including main protective and supplementary bonding conductors must be tested for continuity using a low-resistance ohmmeter.

For main protective bonding there is no single fixed value of resistance above which the conductor would be deemed unsuitable.

Each measured value, if indeed it is measurable for very short lengths, should be compared with the relevant value for a particular conductor length and size. Such values are shown in Table 3.2.

Where a supplementary bonding conductor has been installed between *simultaneously accessible* exposed and extraneous conductive parts, because circuit disconnection times cannot be met, then the resistance

Table 3.2   Resistance (W) of Copper Conductors at 20°C

CSA (mm²)	Length (m)									
	5	10	15	20	25	30	35	40	45	50
1	0.09	0.18	0.27	0.36	0.45	0.54	0.63	0.72	0.82	0.9
1.5	0.06	0.12	0.18	0.24	0.3	0.36	0.43	0.48	0.55	0.6
2.5	0.04	0.07	0.11	0.15	0.19	0.22	0.26	0.03	0.33	0.37
4	0.023	0.05	0.07	0.09	0.12	0.14	0.16	0.18	0.21	0.23
6	0.02	0.03	0.05	0.06	0.08	0.09	0.11	0.13	0.14	0.16
10	0.01	0.02	0.03	0.04	0.05	0.06	0.063	0.07	0.08	0.09
16	0.006	0.01	0.02	0.023	0.03	0.034	0.04	0.05	0.05	0.06
25	0.004	0.007	0.01	0.015	0.02	0.022	0.026	0.03	0.033	0.04
35	0.003	0.005	0.008	0.01	0.013	0.016	0.019	0.02	0.024	0.03

(R) of the conductor, must be equal to or less than $50/I_a$. So, $R \leq 50/I_a$, where 50 is the voltage above which exposed metalwork should not rise, and I_a is the minimum current causing operation of the circuit protective device within 5 s.

For example, suppose a 45 A BS 3036 fuse protects a cooker circuit, the disconnection time for the circuit cannot be met, and so a supplementary bonding conductor has been installed between the cooker case and an adjacent central heating radiator. The resistance (R) of that conductor should not be greater than $50/I_a$, and I_a in this case is 145 A (see Figure 3A2(b) of the IET Regulations), that is

$50/145 = 0.34 \ \Omega$

How then do we conduct a test to establish continuity of main protective or supplementary bonding conductors? Quite simple really, just connect the leads from a low-resistance ohmmeter to the ends of the bonding conductor (Figure 3.2). One end should be disconnected from its bonding clamp, otherwise any measurement may include the resistance of parallel paths of other earthed metal-work. Remember to zero or null the instrument or, if this facility is not available, record the resistance of the test leads so that this value can be subtracted from the test reading.

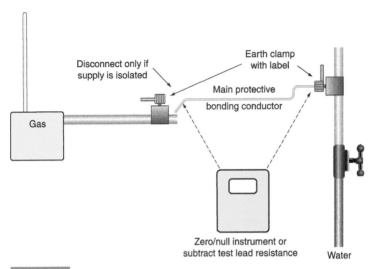

FIGURE 3.2 Testing for continuity.

Important Note

If the installation is in operation, then never disconnect main protective bonding conductors unless the supply can be isolated. Without isolation, persons and livestock are at risk of electric shock.

The continuity of circuit protective conductors (cpcs) may be established in the same way, but a second method is preferred, as the results of this second test indicate the value of $(R_1 + R_2)$ for the circuit in question.

The test is conducted in the following manner:

1. Temporarily link together the line conductor and cpc of the circuit concerned in the distribution board or consumer unit.
2. Test between line and cpc at each outlet in the circuit. A reading indicates continuity.
3. Record the test result obtained at the furthest point in the circuit. This value is $(R_1 + R_2)$ for the circuit.

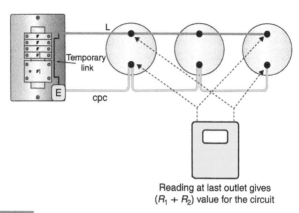

Reading at last outlet gives
$(R_1 + R_2)$ value for the circuit

FIGURE 3.3 Testing cpc continuity.

Figure 3.3 illustrates the above method. There may be some difficulty in determining the $(R_1 + R_2)$ values of circuits in installations that comprise steel conduit and trunking, and/or SWA and MI cables because of the parallel earth paths that are likely to exist. In these cases, continuity tests may have to be carried out at the installation stage before accessories are connected or terminations made off as well as after completion.

Although it is no longer considered good working practice to use steel conduit or trunking as a protective conductor, it is permitted, and hence its continuity must be proved. The enclosure must be inspected along its length to ensure that it is sound and then the standard low-resistance test is performed.

CONTINUITY OF RING FINAL CIRCUIT CONDUCTORS

There are two main reasons for conducting this test:

1. To establish that interconnections in the ring do not exist.
2. To ensure that the cpc is continuous, and indicate the value of $(R_1 + R_2)$ for the ring.

What then are interconnections in a ring circuit, and why is it important to locate them? Figure 3.4 shows a ring final circuit with an interconnection.

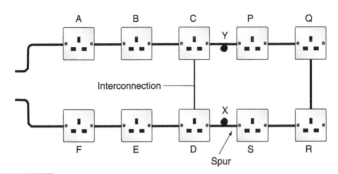

FIGURE 3.4 Ring circuit with interconnection.

The most likely cause of the situation shown in Figure 3.4 is where a DIY enthusiast has added sockets P, Q, R and S to an existing ring A, B, C, D, E and F.

In itself there is nothing wrong with this. The problem arises if a break occurs at, say, point Y, or the terminations fail in socket C or P. Then there would be four sockets all fed from the point X which would then become a spur.

So, how do we identify such a situation with or without breaks at point 'Y'? A simple resistance test between the ends of the line, neutral or cpcs will only indicate that a circuit exists, whether there are interconnections or not. The following test method is based on the philosophy that the resistance measured across any diameter of a perfect circle of conductor will always be the same value (Figure 3.5).

The perfect circle of conductor is achieved by cross connecting the line and neutral legs of the ring (Figure 3.6).

The test procedure is as follows:

1. Identify the opposite legs of the ring. This is quite easy with sheathed cables, but with singles, each conductor will have to be identified, probably by taking resistance measurements between each one and the closest socket outlet. This will give three high readings and three low readings thus establishing the opposite legs.
2. Take a resistance measurement between the ends of each conductor loop. Record this value.

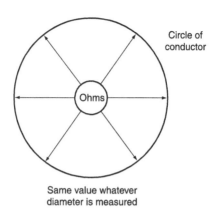

FIGURE 3.5

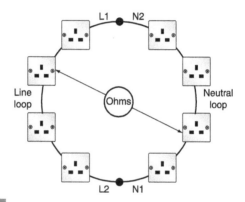

FIGURE 3.6

3. Cross connect the opposite ends of the line and neutral loops (Figure 3.7).

4. Measure between line and neutral at each socket on the ring. The readings obtained should be, for a perfect ring, substantially the same. If an interconnection existed such as shown in Figure 3.4, then sockets A to F would all have similar readings, and those beyond the interconnection would have gradually increasing values to approximately the mid point of the ring, then decreasing values back towards the interconnection. If a break had occurred at point

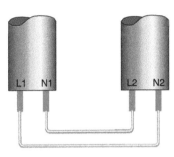

FIGURE 3.7 L and N cross connections.

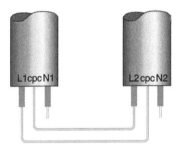

FIGURE 3.8 L and cpc cross connections.

Y then the readings from socket S would increase to a maximum at socket P. One or two high readings are likely to indicate either loose connections or spurs. A null reading (i.e. an open-circuit indication) is probably a reverse polarity, either line- or neutral-cpc reversal. These faults would clearly be rectified and the test at the suspect socket(s) repeated.

5. Repeat the above procedure, but in this case cross connect the line and cpc loops (Figure 3.8).

In this instance, if the cable is of the flat twin type, the readings at each socket will increase very slightly and then decrease around the ring. This difference, due to the line and cpc being different sizes, will not be significant enough to cause any concern. The measured value is very important, it is $R_1 + R_2$ for the ring.

As before, loose connections, spurs and, in this case, L–N cross polarity will be picked up.

Table 3.3

	L1–P2	N1–N2	cpc 1–cpc 2
Initial measurements	0.52	0.52	0.86
Reading at each socket	0.26	0.26	0.32–0.34
For spurs, each metre in length will add the following resistance to the above values	0.015	0.015	0.02

The details in Table 3.3 are typical approximate ohmic values for a healthy 70 m ring final circuit wired in 2.5/1.5 flat twin and cpc cable. (In this case the cpc will be approximately $1.67 \times$ the L or N resistance.)

As already mentioned, null readings may indicate a reverse polarity. They could also indicate twisted conductors not in their terminal housing. The examples shown in Figure 3.9 may help to explain these situations.

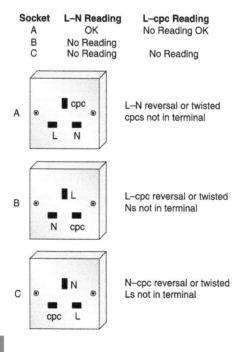

Socket	L–N Reading	L–cpc Reading
A	OK	No Reading OK
B	No Reading	
C	No Reading	No Reading

A — L–N reversal or twisted cpcs not in terminal

B — L–cpc reversal or twisted Ns not in terminal

C — N–cpc reversal or twisted Ls not in terminal

FIGURE 3.9

Socket	L–N Reading	L–cpc Reading
A	OK	No reading
B	No reading	OK
C	No reading	No reading

INSULATION RESISTANCE

This is probably the most used and yet abused test of them all. Affectionately known as 'meggering', an *insulation resistance test* is performed in order to ensure that the insulation of conductors, accessories and equipment is in a healthy condition, and will prevent dangerous leakage currents between conductors and between conductors and earth. It also indicates whether any short circuits exist.

Insulation resistance, as just discussed, is the resistance measured between conductors and is made up of countless millions of resistances in parallel (Figure 3.10).

The more resistances there are in parallel, the *lower* the overall resistance, and in consequence, the longer a cable, the lower the insulation resistance. Add to this the fact that almost all installation circuits are also wired in parallel, and it becomes apparent that tests on large installations may give, if measured as a whole, pessimistically low values, even if there are no faults.

Under these circumstances, it is usual to break down such large installations into smaller sections, floor by floor, distribution circuit (submain), etc. This also helps, in the case of periodic testing, to minimize disruption. The test procedure is as follows:

1. Disconnect all items of equipment such as capacitors and indicator lamps as these are likely to give misleading results. Remove any items of equipment likely to be damaged by the test, such as dimmer switches, electronic timers, etc. Remove all lamps and accessories and disconnect fluorescent and discharge fittings. Ensure that the installation is disconnected from the supply, all fuses are in place, and circuit breakers and switches are in the on position. In some instances it may be impracticable to remove lamps, etc., and in this

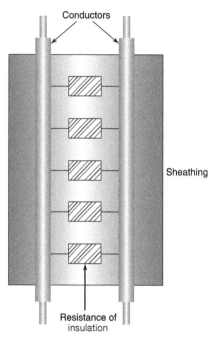

FIGURE 3.10 Resistance of insulation.

case the local switch controlling such equipment may be left in the off position.

2. Join together all live conductors of the supply and test between this join and earth. Alternatively, test between each live conductor and earth in turn.

3. Test between line and neutral. For three-phase systems, join together all lines and test between this join and neutral. Then test between each of the lines. Alternatively, test between each of the live conductors in turn. Installations incorporating two-way lighting systems should be tested twice with the two-way switches in alternative positions.

Table 3.4 gives the test voltages and minimum values of insulation resistance for ELV and LV systems.

Table 3.4

System	Test Voltage	Minimum Insulation Resistance
SELV and PELV	250 V DC	0.5 MΩ
LV up to 500 V	500 V DC	1 MΩ
Over 500 V	1000 V DC	1.0 MΩ

If a value of less than 2 MΩ is recorded it may indicate a situation where a fault is developing, but as yet still complies with the minimum permissible value. In this case each circuit should be tested separately to find the faulty circuit.

Where electronic devices cannot be disconnected, test only between line and neutral connected together and protective earth.

Example 3.1

An installation comprises six circuits with individual insulation resistances of 2.5, 8, 200, 200, 200 and 200 MΩ, and so the total insulation resistance will be:

$$\frac{1}{R_t} = \frac{1}{2.5} + \frac{1}{8} + \frac{1}{200} + \frac{1}{200} + \frac{1}{200} + \frac{1}{200}$$

$$= 0.4 + 0.125 + 0.005 + 0.005 + 0.005 + 0.005$$

$$= 0.545$$

$$R_t = \frac{1}{0.545} = 1.83 \text{ M}\Omega$$

This is clearly greater than the 1 MΩ minimum but less than 2 MΩ, and so an investigation should be conducted to identify the suspect circuit/s.

POLARITY

This simple test, often overlooked, is just as important as all the others, and many serious injuries and electrocutions could have been prevented if only polarity checks had been carried out.

The requirements are:

1. All fuses and single pole switches are in the line conductor.
2. The centre contact of an Edison screw type lampholder is connected to the line conductor (except E 14 and E 27, which are new types with an insulating thread material).
3. All socket outlets and similar accessories are correctly wired.

Although polarity is towards the end of the recommended test sequence, it would seem sensible, on lighting circuits, for example, to conduct this test at the same time as that for continuity of cpcs (Figure 3.11).

As discussed earlier, polarity on ring final circuit conductors is achieved simply by conducting the ring circuit test. For radial socket-outlet circuits, however, this is a little more difficult. The continuity of the cpc will have already been proved by linking line and cpc and measuring between the same terminals at each socket. Whilst a line–cpc reversal would not have shown, a line–neutral reversal would, as there would have been no reading at the socket in question. This would have been remedied, and so only line-cpc reversals need to be checked. This can be done by linking together line and neutral at the origin and testing between the same terminals at each socket. A line–cpc reversal will result in no reading at the socket in question.

When the supply is connected, it is important to check that the incoming supply is correct. This is done using an approved voltage indicator at the intake position or close to it. In the case of a three-phase supply a phase sequence indicator would be used.

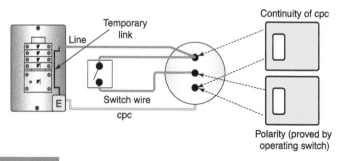

FIGURE 3.11 Lighting polarity.

EARTH FAULT LOOP IMPEDANCE

This is very important but, sadly, poorly understood. So let us remind ourselves of the component parts of the earth fault loop path (Figure 3.12). Starting at the point of fault:

1. The cpc.
2. The earthing conductor and earthing terminal.
3. The return path via the earth for TT systems, and the metallic return path in the case of TN-S or TN-C-S systems. In the latter case the metallic return is the PEN conductor.
4. The earthed neutral of the supply transformer.
5. The transformer winding.
6. The line conductor back to the point of fault.

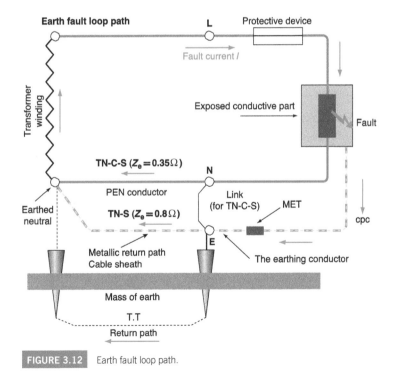

FIGURE 3.12 Earth fault loop path.

Overcurrent protective devices must, under earth fault conditions, disconnect fast enough to reduce the risk of electric shock. This is achieved if the actual value of the earth fault loop impedance does not exceed the tabulated maximum values given in the IET Regulations.

The purpose of the test, therefore, is to determine the actual value of the loop impedance (Z_s), for comparison with those maximum values, and it is conducted as follows:

1. Ensure that all main protective bonding conductors are in place.
2. Connect the test instrument either by its BS 1363 plug, or the 'flying leads', to the line, neutral and earth terminals at the remote end of the circuit under test. (If a neutral is not available, e.g. in the case of a three-phase motor, connect the neutral probe to earth.)
3. Press to test and record the value indicated.

It must be understood that this instrument reading is *not valid for direct comparison with the tabulated maximum values*, as account must be taken of the ambient temperature at the time of test, and the maximum conductor operating temperature, both of which will have an effect on conductor resistance. Hence, the $(R_1 + R_2)$ could be greater at the time of fault than at the time of test.

So, our measured value of Z_s must be corrected to allow for these possible increases in temperature occurring at a later date. This requires actually measuring the ambient temperature and applying factors in a formula.

Clearly, this method of correcting Z_s is time consuming and unlikely to be commonly used. Hence, a rule of thumb method may be applied which simply requires that the measured value of Z_s does not exceed 0.8 of the appropriate tabulated value. Table 3.6 gives the 0.8 values of tabulated loop impedance for direct comparison with measured values.

In effect, a loop impedance test places a line/earth fault on the installation, and if an RCD is present it may not be possible to conduct the test as the device will trip out each time the loop impedance tester button is pressed. Unless the instrument is of a type that has a built-in guard against such tripping, the value of Z_s will have to be determined from measured values of Z_e and $(R_1 + R_2)$, and the 0.8 rule applied.

Important Note

Never short out an RCD in order to conduct this test.

As a loop impedance test creates a high earth fault current, albeit for a short space of time, some lower rated circuit breakers may operate, resulting in the same situation as with an RCD, and Z_s will have to be calculated. It is not really good practice temporarily to replace the circuit breaker with one of a higher rating.

External loop impedance Z_e

The value of Z_e is measured at the intake position on the supply side and with all main protective bonding conductors disconnected. Unless the installation can be isolated from the supply, this test should not be carried out, as a potential shock risk will exist with the supply on and the main bonding disconnected.

Prospective fault current

This would normally be carried out at the same time as the measurement for Z_e using a PFC or PSCC (prospective short-circuit current) tester. If this value cannot be measured it must be ascertained by either enquiry or calculation.

Note

Never short out an RCD in order to conduct this test. In domestic premises with consumer units to BS EN 61439-3 and a Distribution Network Operator declared value of 16kA, then no measurement or calculation is required

Be sure to follow the instrument manufacturer's instructions. Measure Ipf line to neutral, if three leads, place N/protective conductors on neutral bar (for typical instruments). Record highest reading and determine a value for three-phase where Ipf (short circuit current) is approximately 2×3 single-phase value.

Table 3.5 Corrected Maximum Z_s Values for Comparison with Measured Values

PROTECTIVE DEVICE		PROTECTION RATING																		
		5	6	10	15	16	20	25	30	32	40	45	50	60	63	80	100	125	160	200
BS EN 60898 and 61009 type **B**	0.4s & 5s		5.82	3.5		2.19	1.75	1.4		1.09	0.87		0.7		0.56	0.5	0.35	0.29		
BS EN 60898 and 61009 type **C**	0.4s & 5s		2.9	1.75		1.09	0.87	0.71		0.54	0.44		0.35		0.28	0.2	0.17	0.14		
BS EN 60898 and 61009 type **D**	0.4s		1.46	0.87		0.54	0.44	0.35		0.28	0.22		0.17		0.13	0.1	0.08	0.07		
	5s		2.91	1.75		1.09	0.87	0.69		0.55	0.44		0.35		0.27	0.22	0.17	0.14		
BS 3036 Semi-enclosed	0.4s	7.22			1.94		1.34		0.83			0.45		0.32						
	5s	13.45			4.1		2.9		2			1.2		0.85			0.4			
BS 1362 Cartridge	0.4s	12.5 *(3A*)*			1.8 *(13A*)*															
	5s	17.63			2.9															

BS 88-2 Bolted type E and Clip-in type G

Time														
0.4s	6.24	3.71		1.95	1.35	1.03	0.79	0.6		0.46		0.35		
5s	9.73	5.46	3.17	2.24	1.75	1.4	1.03	0.79	0.64	0.4	0.35	0.26	0.21	0.2

BS 88-3 Cartridge type C

Time											
0.4s	7.94			1.84	1.55		0.73		0.46		0.35
5s	11.62			2.14	2.57		1.24	0.79	0.55	0.4	0.3

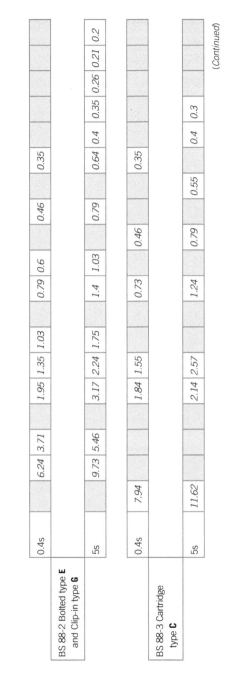

(Continued)

Table 3.6 Corrected Maximum Z_s Values for Comparison with Measured Values

OBSOLETE FUSES & CIRCUIT BREAKERS

PROTECTION RATING

MAXIMUM CORRECTED Zs VALUES

	Time	5	6	10	15	16	20	25	30	32	40	45	50	60	63	80	100	125	160	200
BS 88 – 2.2 & 88-6	0.4s		6.48	3.89		2.05	1.35	1.09		0.79										
	5s		10.26	5.64		3.16	2.2	1.75		1.4	1.03		0.79		0.64	0.5	0.32	0.25	0.19	0.2
BS 1361 Cartridge	0.4s	7.94			2.45		1.29		0.87											
	5s	12.46			3.8		2.13		1.4			0.75		0.53		0.4	0.28			
BS 3871 type 1	0.4s & 5s	8.74	7.22	4.37	2.9	2.73	2.19	1.75	1.45	1.37	1.09	0.97	0.87		0.69					
BS 3871 type 2	0.4s & 5s	5	4.15	2.49	1.66	1.56	1.24	1	0.83	0.78	0.64	0.55	0.49		0.42					
BS 3871 type 3	0.4s & 5s	3.5	2.85	1.75	1.16	1.09	0.87	0.7	0.58	0.54	0.44	0.39	0.35		0.28					

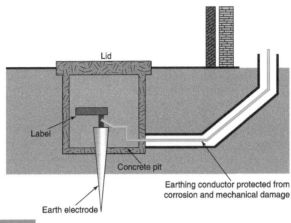

FIGURE 3.13 Earth electrode installation.

EARTH ELECTRODE RESISTANCE

In many rural areas, the supply system is TT and hence reliance is placed on the general mass of earth for a return path under earth fault conditions. Connection to earth is made by an electrode, usually of the rod type, and preferably installed as shown in Figure 3.13.

In order to determine the resistance of the earth return path, it is necessary to measure the resistance that the electrode has with earth. If we were to make such measurements at increasingly longer distances from the electrode, we would notice an increase in resistance up to about 2.5–3 m from the rod, after which no further increase in resistance would be noticed (Figure 3.14).

The maximum resistance recorded is the electrode resistance and the area that extends the 2.5–3 m beyond the electrode is known as the earth electrode resistance area.

There are two methods of making the measurement, one using a proprietary instrument and the other using a loop impedance tester.

Method 1: Usually where no AC supply is available

This method is based on the principle of the potential divider (Figure 3.15).

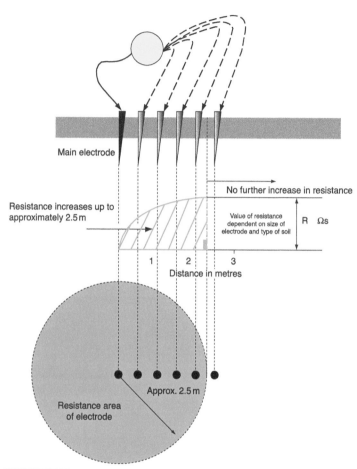

Main electrode

No further increase in resistance

Resistance increases up to approximately 2.5 m

Value of resistance dependent on size of electrode and type of soil

R   Ωs

1   2   3
Distance in metres

Approx. 2.5 m

Resistance area of electrode

FIGURE 3.14   Earth electrode resistance area.

By varying the position of the slider the resistance at any point may be calculated from $R = V/I$.

The earth electrode resistance test is conducted in a similar fashion, with the earth replacing the resistance and a potential electrode replacing the slider (Figure 3.16). In Figure 3.16 the earthing conductor to the electrode under test is temporarily disconnected.

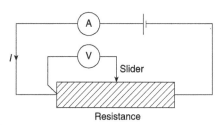

FIGURE 3.15 Potential divider.

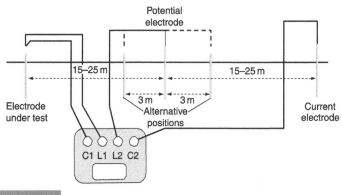

FIGURE 3.16 Earth electrode test.

The method of test is as follows:

1. Place the current electrode (C2) away from the electrode under test, approximately 10 times its length (i.e. 30 m for a 3 m rod).
2. Place the potential electrode mid way.
3. Connect test instrument as shown.
4. Record resistance value.
5. Move the potential electrode approximately 3 m either side of the mid position, and record these two readings.
6. Take an average of these three readings (this is the earth electrode resistance).
7. Determine the maximum deviation or difference of this average from the three readings.

8. Express this deviation as a percentage of the average reading.
9. Multiply this percentage deviation by 1.2.
10. Provided this value does not exceed a figure of 5% then the accuracy of the measurement is considered acceptable.

If three readings obtained from an earth electrode resistance test were 181, 185 and 179 Ω, what is the value of the electrode resistance and is the accuracy of the measurement acceptable?

$$\text{Average value} = \frac{181 + 185 + 179}{3}$$

$$\text{Maximum deviation} = 185 - 181.67$$
$$= 3.33$$

$$\text{Expressed as a percentage of the average} = \frac{3.33 \times 100}{181.67}$$
$$= 1.83\%$$

$$\text{Measurement accuracy} = 1.83\% \times 1.2$$
$$= 2.2\% \text{ (which is acceptable)}$$

Method 2: For TT systems protected by an RCD

In this case, an earth fault loop impedance test is carried out between the incoming line terminal and the electrode (a standard test for Z_e).

The value obtained is added to the cpc resistance of the protected circuits and this value is multiplied by the operating current of the RCD. The

Table 3.7		
RCD Type	**Half Rated**	**Full Trip Current**
BS 4239 and BS 7288 sockets	No trip	< 200 ms
BS 4239 with time delay	No trip	$\frac{1}{2}$time delay +200 ms to time delay +200 ms
BS EN 61008 or BS EN 61009 RCBO	No trip	< 300 ms
As above but Type S with time delay	No trip	130 < to < 500 ms

resulting value should not exceed 50 V. If it does, then Method 1 should be used to check the actual value of the electrode resistance.

FUNCTIONAL TESTING
RCD/RCBO operation

Where RCDs/RCBOs are fitted, it is essential that they operate within set parameters. The RCD testers used are designed to do just this, and the basic tests required are as follows:

1. Set the test instrument to the rating of the RCD.
2. Set the test instrument to half rated trip.
3. Operate the instrument and the RCD should not trip.
4. Set the instrument to deliver the full rated tripping current of the RCD.
5. Operate the instrument and the RCD should trip out in the required time.

There seems to be a popular misconception regarding the ratings and uses of RCDs in that they are the panacea for all electrical ills and the only useful rating is 30 mA!

Firstly, RCDs are not fail safe devices, they are electromechanical in operation and can malfunction, and secondly, RCDs are manufactured in ratings from 5 to 500 mA and have many uses. Table 3.8 illustrates RCD ratings and applications.

For 30 mA ratings, it is required that the RCD be injected with a current five times its operating current and the tripping time should not exceed 40 ms.

Where loop impedance values cannot be met, RCDs of an appropriate rating can be installed. Their rating can be determined from

$$I_{\Delta n} = 50/Z_s$$

where:

$I_{\Delta n}$ is the rated operating current of the device
50 is the touch voltage
Z_s is the measured loop impedance

Table 3.8	Requirements for RCD Protection

30 mA
- All socket outlets rated at not more than 32A.
- All circuits supplying luminaires in domestic premises.
- Mobile equipment rated at not more than 32 A for use outdoors.
- All circuits in a bath/shower room.
- Preferred for all circuits in a TT system.
- All cables installed less than 50 mm from the surface of a wall or partition unless they are adequately mechanically protected or have earthed screening or armouring, and also at any depth if the construction of the wall or partition includes metallic parts.
- In zones 0, 1 and 2 of swimming pool locations.
- All circuits in a location containing saunas, etc.
- Socket outlet final circuits not exceeding 32 A in agricultural locations.
- Circuits supplying Class II equipment in restrictive conductive locations.
- Each socket outlet in caravan parks and marinas and final circuit for houseboats.
- All socket outlet circuits rated not more than 32 A for show stands, etc.
- All socket outlet circuits rated not more than 32 A for construction sites (where reduced low voltage, etc., is not used).
- All socket outlets supplying equipment outside mobile or transportable units.
- All circuits in caravans.
- All circuits in circuses, etc.
- A circuit supplying Class II heating equipment for floor and ceiling heating systems.

500 mA
- Any circuit supplying one or more socket outlets of rating exceeding 32 A, on a construction site.

300 mA
- At the origin of a temporary supply to circuses, etc.
- Where there is a risk of fire due to storage of combustible materials.
- All circuits (except socket outlets) in agricultural locations.

100 mA
- Socket outlets of rating exceeding 32 A in agricultural locations.
Where loop impedances are too high, RCD ratings can be calculated.

All RCDs have a built-in test facility in the form of a test button. Operating this test facility creates an artificial out-of-balance condition that causes the device to trip. This only checks the mechanics of the tripping operation, it is not a substitute for the tests just discussed.

All other items of equipment such as switchgear, controlgear interlocks, etc., must be checked to ensure that they are correctly mounted and adjusted and that they function correctly.

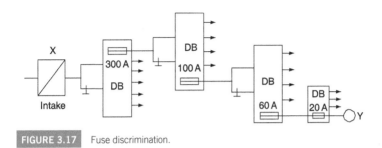

FIGURE 3.17 Fuse discrimination.

PROSPECTIVE FAULT CURRENT

If we are able to remove all the protective devices in a system and connect a shorting link between line and neutral conductors, the resulting current that would flow at the point where the link was placed is known as the PFC or in this case PSCC. This is all very interesting, you may comment, but why would one need to know all this information?

When a short circuit occurs, the device protecting the circuit in question has to open that circuit. If the device is incorrectly selected, a violent and possibly damaging explosion could result.

We have probably all experienced the loud bang and resulting blackened and molten copper splattered fuse holders that often occur with rewireable fuses.

Each protective device has a rated breaking capacity, which indicates the level of fault current that it can interrupt without arcing or scattering hot particles or damaging associated conductors. For example, rewireable fuses to BS 3036 ranging between 5 and 200 A have breaking capacities ranging from 1000 to 12 000 A, and HRC BS 88 types from 10 000 to 80 000 A. MCBs to BS 3871 (now obsolete but still in use) are designed for one fault level irrespective of size, and have ranges of 1000 A (M1), 1500 A (M1.5), 3000 A (M3), 4500 A (M4.5), 6000 A (M6) and 9000 A (M9). Circuit breakers to BS EN 60898 (European standard) have two ratings: Icn and Ics. Icn is the rating of fault current after which the breaker should not be used, and Ics is the fault rating after which the breaker is assumed to be able to be used without loss of performance.

So it is important to ensure that the breaking capacity of a protective device is capable of interrupting at least the PFC at the point at which it is installed; hence the need to measure this current.

The type of instrument most commonly used is a dual device which also measures earth fault loop impedance. It is simply connected or plugged into the circuit close to the point of protection, a button pushed and the PFC value read.

It is not always necessary to test at every point at which a protective device is installed, if the breaking capacity of the lowest rated device in the circuit is greater than the PSCC at the origin. In Figure 3.17, clearly if the breaking capacity of the 20 A protective device at Y is greater than the PSCC at X, then there is no need to test at the other DBs as the protection there will have even greater breaking capacities than the 20 A device.

CHAPTER 4

Fault Finding

Important terms/topics covered in this chapter:

- Signs and symptoms of faults
- Typical faults and their diagnosis

At the end of this chapter the reader should:

- have an understanding of typical faults on a range of systems,
- know the correct approach to fault diagnosis.

Starting this chapter has been as difficult as finding a fault on a complex system. The old adage of 'start at the beginning and finish at the end' is all right for storytellers or authors of technical books, but is not always sensible for fault finders! Perhaps it would be wise to begin by determining what exactly a fault is.

A fault is probably best defined as a disturbance in an electrical system of such magnitude as to cause a malfunction of that system. It must be remembered, of course, that such disturbances may be the secondary effects of mechanical damage or equipment failure. Actual electrical faults, or should one say faults that are caused by 'electricity', are rare and are confined in the main to bad design and/or installation or deterioration and ageing.

However, all of this still does little to indicate where to begin looking for a fault. Only experience will allow one to pinpoint the exact seat of a breakdown, but a logical approach to fault finding will save a great deal of time and frustration.

Important Note

The Electricity at Work Regulations prohibit work on live systems unless it is unreasonable in all circumstances to work on them de-energised, so, if fault finding can be carried out with circuits switched off ensure safe isolation is carried out. (See Appendix 1)

SIGNS AND SYMPTOMS

Always look for tell-tale signs that may indicate what kind of fault has caused the system breakdown. If possible, ask persons present to describe the events that led up to the fault. For example, inspection of a rewireable fuse may reveal one of the conditions shown in Figures 4.1 and 4.2.

The IET Wiring Regulations define overload and short-circuit current as follows:

- **Overload current**: An overcurrent occurring in a circuit which is electrically sound.
- **Short-circuit current**: An overcurrent resulting from a fault of negligible impedance between live conductors having a difference in potential under normal operating conditions.

It is likely that other personnel will have replaced fuses or reset circuit breakers to no avail before you, the tester, is called in. In these circumstances they will be in a position to indicate that protective devices operate immediately or after a delay. This at least will give some clue to the type of fault.

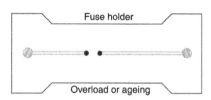

FIGURE 4.1

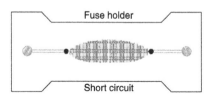

FIGURE 4.2

Let us now consider typical faults on some of the systems already discussed in this book, and endeavour to suggest some solutions.

Remember, however, that it is impossible to catalogue every fault and its cause that may occur in every system.

Diagnostic flow-charts have been included for some of the following faults.

RING AND RADIAL SOCKET OUTLET CIRCUITS

Reported fault

'My fuse/circuit breaker blows at odd unspecified times during the day, with no particular appliance plugged into any particular socket.'

Diagnosis

1. This could be incorrect fusing, but that is unlikely because then the fault would always have been present. It is probably an insulation breakdown.
2. Conduct an insulation resistance test (remember to remove all appliances). If the reading is low, then:
3. Go to the centre of the circuit and disconnect the socket; disconnect the ends in the fuse board if the circuit is a ring; and test in both directions.
4. Probably only one side will indicate a fault, so subdivide. Keep testing and subdividing until the faulty cable section or socket is located (Figure 4.3).

New ring final circuit installations

In Chapter 3 three methods are described for testing the continuity of the final conductors. The purpose is to locate interconnections in the ring. Should such an interconnection prove to exist, its position in the ring must be located. Methods 1 and 2 will indicate a fault, but location is achieved by systematically removing sockets to find the interconnection. Method 3 will give an indication of the location of the fault because tests on sockets nearest to the fuse board will give similar readings, whereas those beyond the interconnection will be considerably different.

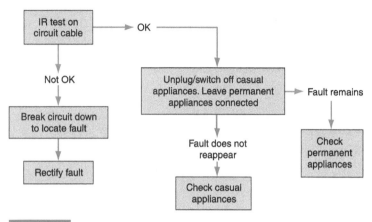

FIGURE 4.3 Socket outlet fault flow chart.

RADIAL CIRCUITS FEEDING FIXED EQUIPMENT

Reported fault

'Every time I start my hydraulic press, the fuse blows in the board.'

Diagnosis

1. This is probably a fault on the press motor windings or the starter. A supply cable fault would normally operate the protection without an attempt to start the motor.
2. Check the starter for obvious signs of damage, burning, etc. If it seems all right, then:
3. Do an insulation resistance test on the motor windings (Figure 4.4).

Reported fault

'The circuit breaker that protects my roof extraction fan has tripped out and will not reset even with the fan switched off.'

Diagnosis

This is almost certainly a cable fault. Do an insulation resistance test. If a fault is indicated, start looking!

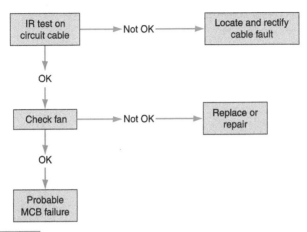

FIGURE 4.4 Fixed equipment fault flow chart.

CABLE FAULT LOCATION

The comment 'start looking' in the last example is quite acceptable for shortish, accessible cable runs and where some sign of damage is evident. However, for longer, hidden cable routes, especially under ground, visual location is impossible. In these instances special tests must be applied, such as the following.

Murray loop test

This test and its variations are based on the principle of the Wheatstone bridge. It is used for the location of short-circuit faults. In Figure 4.5, when the variable resistances are adjusted such that the reading on the ammeter is zero, it will be found that

$A/B = C/D$

Let us now replace resistors C and D with the same cores of a faulty cable (Figure 4.6). The ratios will be the same, as resistance is proportional to length. A link is placed between a sound core and a faulty core. It will be seen that now C is replaced by $L + (L - X)$, and D by X, so that a greater accuracy can be achieved if the test is then repeated at the

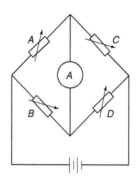

FIGURE 4.5 Wheatstone bridge.

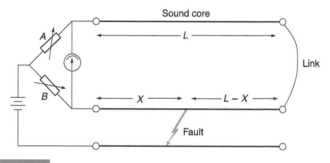

FIGURE 4.6 Murray loop test.

other end of the cable and the mean position of X is taken as the point of fault.

$$A/B = [L + (L - X)]/X$$

Hence

$$A/B = (2L - X)/X$$
$$AX = 2LB - BX$$
$$AX + BX = 2LB$$
$$X(A + B) = 2LB$$
$$X = \frac{2LB}{(A + B)}$$

Variable resistors *A* and *B* are usually incorporated in a single bridge instrument, and the resistances are adjusted for a zero reading.

EMERGENCY LIGHTING

There is not a lot to go wrong with these systems other than the cable faults and internal failure of the unit. Cable faults are located in the way previously described; however, do not conduct insulation resistance tests with the unit still connected, or the internal electronic circuitry may be damaged.

Individual units have LEDs that indicate a healthy state of charge. If this is out,

1. Check the mains supply. If it is all right, then:
2. Check the output of the charging unit at the battery terminals. If this is all right, check the battery separately.

If the LED is lit but the emergency light does not function when the mains supply is removed, check the lamps. If these are functioning, the circuitry is probably faulty.

SECURITY AND FIRE ALARM SYSTEMS

Fire alarm systems have to be inspected at regular intervals and hence tend to be very reliable. The most common faults in such systems are false operation of heat or smoke detectors, perhaps owing to a change in the use of the protected area since they were first installed, or a break in the sensor cable. As the continuity of these cables is monitored, any fault will bring up a fault condition on the main panel. Simple continuity tests on cable between sensors should reveal the broken section.

Security systems are somewhat similar in that faults on cables or sensors will bring up an alarm condition on the main panel. Remove the ends of the zone in question and insert a shorting link. If the alarm still activates when switched on, a fault in the internal circuitry is indicated. If not, start at the middle of a zone and work back to the panel, similar to the procedure used on ring/radial circuits, until the faulty cable or sensor is found.

CALL SYSTEMS

The faults on these systems, once again, are broadly similar to those on alarm systems.

Reported fault

'The patient in room 42 pushes the call button but nothing happens.'

Diagnosis

This is either a faulty push-button, a break in the cable, or a defective relay or electronic component. Trial and error is the rule here: check out the simple and obvious first before taking the control panel apart (Figure 4.7).

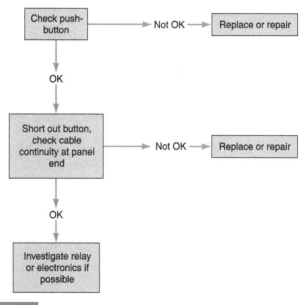

FIGURE 4.7   Call system fault flow chart.

CENTRAL HEATING SYSTEMS

The most difficult faults to locate are those that occur during the commissioning of a newly installed system, and are almost always due to incorrect wiring. Naturally, not all the faults that could occur on all the available systems can be detailed, but there are one or two common mistakes that are worth mentioning:

1. The CH is selected but the HW comes on, and vice versa. This is simply a reversal of the CH and HW feeds in the programmer.
2. On a mid-position valve system, a reversal of the leads to the demand and satisfy terminals in the cylinder thermostat will cause various incorrect sequences depending on which heating options have been chosen.

On one occasion, the author installed the control wiring for a CH system, and was perplexed when the room thermostat operated the HW and the cylinder thermostat operated the CH. Much tearing of hair followed, as the system wiring proved to be correct. Some time later it was discovered that the DIY householder had installed the mid-position valve back-to-front.

With existing systems, faults tend to be confined to equipment failure such as defective programmers, valves and pumps. Once again it must be stressed that fault location on such systems is eased considerably when the system is fully understood, and this generally comes with experience.

MOTOR STARTER CIRCUITS

The usual cable faults we have already discussed must of course be considered, but faults in the starter itself are usually confined to coil failure or contact deterioration.

Coil failure

If this should occur the motor will not start, for the obvious reason that the control circuit is defective.

Reported fault

'Everything was all right this morning – the motor was running – but this afternoon everything just stopped, and will not start up again.'

Diagnosis

Could be operation of circuit protection, coil failure or overload operation due to a motor fault or excessive mechanical loading.

1. Check overloads. If they have operated, carry out insulation resistance test on motor windings. If all OK, then:
2. Check circuit protection. If all OK, then fault is probably coil failure. This can be verified by testing the coil, or holding in the contactor manually (Figure 4.8).

Contact deterioration

Here is one of the few genuine faults caused by the effects of 'electricity'. Over time, constant operation with its attendant arcing will burn and pit the surface of contacts. This can lead to poor connection between the faces, or even the complete destruction of the contacts.

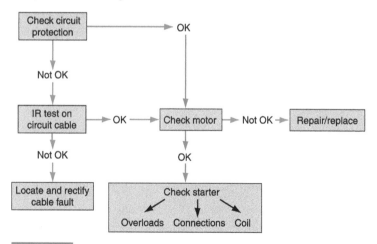

FIGURE 4.8 Motor circuit fault flow chart.

Reported fault

'We shut down the conveyor for lunch, and when we try to start it up again, it only works if we keep the start button in.'

Diagnosis

1. This is clearly a 'hold-on' circuit fault. The coil must be OK or the conveyor would not work.
2. Check the continuity across the hold-on contacts with the contactor depressed manually (supply isolated of course). If open circuit is found, strip the contactor and investigate.

Protection faults

Other causes of motor circuit failure often result from incorrect protection – fuses, MCBs or overloads – or incorrect reconnection after motor replacement.

Motors may take as much as eight times their normal rated current on starting. If the protection is of the wrong type or size, or is set incorrectly, it will operate.

Fuses

Use a motor rated fuse, usually BS 88. These are easily recognizable, as the rating has a prefix M (e.g. M20 is a 20 A motor rated fuse). They are designed to cater for high starting currents.

MCBs/CBs

These range from the now obsolete BS 3871 types 1, 2 and 3 to the latest BS EN 60898 types A, B, C and D. Types 1, 2, A and B are usually too sensitive to handle starting currents, so types 3, C and D are preferred.

Overloads

These are adjustable and their ratings must be suitable for the full load of the motor. Ensure that the overload has been selected to the correct rating.

Motor replacement

If a motor has been diagnosed as being faulty, and is either repaired or replaced, always ensure that the reconnection to the motor terminals is correct. A reversal of any two of the connections to a three-phase direct-on-line motor will reverse its direction of rotation, and a reversal of the connections to the windings of a star-delta motor could have serious implications for the motor together with operation of protective devices.

Figures 4.9 and 4.10 show the terminal arrangements for a six terminal motor and the correct winding connections.

If the connections to any winding are reversed, the magnet fields will work against each other and a serious overload will occur, especially when the starter changes the motor windings to the delta configuration. For example, never allow A2 to be connected to B2 and A1 to C1.

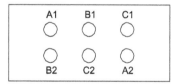

Motor terminal block

FIGURE 4.9

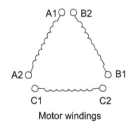

Motor windings

FIGURE 4.10

SOLAR PV SYSTEMS

Problems with solar PV systems are usually rectified during the initial commissioning process and if installed by an MCS accredited contractor, such systems should be relatively fault free.

The most commonly reported fault is that the system is under performing and in many instances this is just an over optimistic expectation of financial returns.

However, there are some instances where underperformance is due to system faults.

Module faults:

- Broken panel glass
- Water ingress
- Damaged cell/s due to excessive shading, e.g. tree growth that was not present at time of installation, or dirt collection
- Cell deterioration due to age.

Cable faults:

- Broken or damaged cables due to wind or snow loading, or damage by rodents.

Loss of performance due to any of the above can be confirmed visually or by checking the MMP voltage. This is found from the module specification and multiplying this figure by the number of modules on a string. On a sunny day the measured MMP voltage should be in the region of 5% of the calculated value.

Inverter faults:

- Most inverters have condition displays and will indicate any faults. Contact the manufacturer
- Overheating due to the location. Inverters should be in a cool place and not restricted in a way to prevent heat dissipation
- Inverters should have been matched correctly with the arrays during the design stage but sometimes an overvoltage or undervoltage may cause them to trip. Lightning is an overvoltage
- If the overvoltage or undervoltage is due to the electricity supply and the inverter regularly trips, then the DNO should be contacted.

CONCLUSION AND A CAUTIONARY TALE

Finally, it must be reiterated that successful and efficient fault finding has its basis in experience. However, for the beginner, use a logical and methodical approach, ask questions of the customer or client, and use some common sense. This in itself will often provide the clues for the diagnosis of system failure.

Once upon a time there was a young electrical apprentice (who shall remain nameless), who was doing a little wiring job (not private, of course). The customer asked the apprentice, while he was there, to investigate whether a cracked socket needed replacing. Of course he agreed to (thinking: 'extras!'), and proceeded to undo the socket without isolating. There was a pop and a flash and the ring circuit fuse operated. 'Silly me,' he thought, 'I'll just repair the fuse, I must have shorted something out in the socket.' 'Hold on,' thought he, 'do this properly: never replace a fuse until the circuit has been checked first.'

An insulation resistance test seemed to be the order of the day, so naturally all appliances were removed from the ring and the test applied. It showed a line neutral short circuit. 'Oh! bother' (or some such expletive), he muttered, 'Where do I start?'

Unfortunately the ring circuit was wired in the form of junction boxes in the upstairs floor void with single drops down to each socket. Splitting the ring meant gaining access to those JBs. Needless to say there were fitted carpets and tongued and grooved floorboards, and no access traps.

Five hours later, with carpets and floorboards up everywhere and a puzzled customer who did not understand such an upheaval to replace a cracked socket, the faulty drop was found. The apprentice ran downstairs but could not immediately find the associated socket outlet. It was there of course, in a cupboard, and next to it was a fridge with a door like a cupboard door, and naturally the fridge was plugged into the socket. 'Heavens above!' and 'flip me' said the apprentice, 'I wish I had asked the customer if there were any appliances I had missed. Worse than that, how do I explain this one away?' Well, needless to say, he did, using the most impressive technical jargon, and vowed never to make the same mistake again.

Shock Risk and Safe Isolation

ELECTRIC SHOCK

This is the passage of current through the body of such magnitude as to have significant harmful effects. Table A1.1 and Figure A1.1 illustrate the generally accepted effects of current passing through the human body. How then are we at risk of electric shock and how do we protect against it?

Table A1.1 Effects of Current Passing Through the Human Body

1–2 mA	Barely perceptible, no harmful effects
5–10 mA	Throw off, painful sensation
10–15 mA	Muscular contraction, can't let go
20–30 mA	Impaired breathing
50 mA and above	Ventricular fibrillation and death

There are two ways in which we can be at risk:

1. Intentional or accidental contact with live parts of equipment or systems that are intended to be live.
2. Contact with conductive parts which are not meant to be live, but which have become live due to a fault.

The conductive parts associated with point (2) above can either be metalwork of electrical equipment and accessories (Class l) and that of electrical wiring systems (e.g. metal conduit and trunking), called exposed conductive parts, or other metalwork (e.g. pipes, radiators and girders), called extraneous conductive parts.

Basic protection

How can we prevent danger to persons and livestock from contact with intentionally live parts? Clearly, we must minimize the risk of such contact and this can be achieved by basic protection, that is:

- insulating any live parts
- ensuring any uninsulated live parts are housed in suitable enclosures and/or are behind barriers.

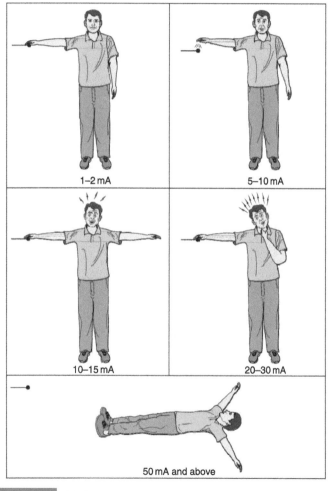

1–2 mA	5–10 mA
10–15 mA	20–30 mA
50 mA and above	

FIGURE A1.1 Shock levels.

The use of a residual current device (RCD) cannot prevent contact with live parts, but it can be used as an addition to any of the other measures taken, provided that its rating, $I_{\Delta n}$, is 30 mA or less.

It should be noted that RCDs must not be used as the sole means of protection.

Fault protection

How can we protect against shock from contact with unintentionally live, exposed or extraneous conductive parts whilst touching earth, or from contact between live exposed and/or extraneous conductive parts? The most common method is by protective earthing, in case of fault and protective equipotential bonding and automatic disconnection in case of a fault.

All extraneous conductive parts are joined together with a main protective bonding conductor and connected to the main earthing terminal, and all exposed conductive parts are connected to the main earthing terminal by the circuit protective conductors. Add to this, overcurrent protection that will operate fast enough when a fault occurs and the risk of severe electric shock is significantly reduced.

What is earth and why and how do we connect to it?

The thin layer of material which covers our planet – rock, clay, chalk or whatever – is what we in the world of electricity refer to as earth. So, why do we need to connect anything to it? After all, it is not as if earth is a good conductor.

It might be wise at this stage to investigate potential difference (PD). A PD is exactly what it says it is: a difference in potential (volts). In this way, two conductors having PDs of, say, 20 and 26 V have a PD between them of $26 - 20 = 6$ V. The original PDs (i.e. 20 and 26 V) are the PDs between 20 V and 0 V and 26 V and 0 V. So where does this 0 V or zero potential come from? The simple answer is, in our case, the earth. The definition of earth is, therefore, the conductive mass of earth, whose electric potential at any point is conventionally taken as zero.

Thus, if we connect a voltmeter between a live part (e.g. the line conductor of a socket outlet) and earth, we may read 230 V; the conductor is at 230 V and the earth at zero. The earth provides a path to complete the circuit. We would measure nothing at all if we connected our voltmeter between, say, the positive 12 V terminal of a car battery and earth, as in this case the earth plays no part in any circuit.

Figure A1.2 illustrates this difference.

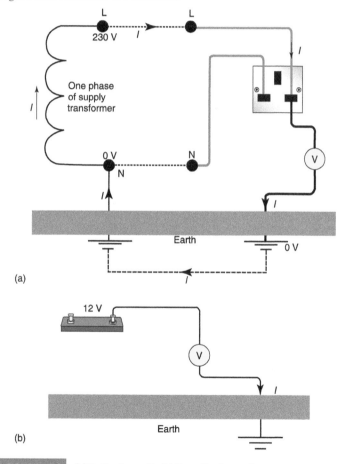

FIGURE A1.2 (a) Earth return path, (b) No earth return path.

So, a person in an installation touching a live part whilst standing on the earth would take the place of the voltmeter and could suffer a severe electric shock. Remember that the accepted lethal level of shock current passing through a person is only 50 mA or 1/20 A.

The same situation would arise if the person were touching a faulty appliance and a gas or water pipe (Figure A1.3).

One method of providing some measure of protection against these effects is, as we have seen, to join together (bond) all metallic parts and connect them to earth. This ensures that all metalwork in a healthy installation is at or near 0 V and, under fault conditions, all metalwork will rise to a similar potential. So, simultaneous contact with two such metal parts would not result in a dangerous shock, as there would be no significant PD between them.

Unfortunately, as mentioned, earth itself is not a good conductor, unless it is very wet. Therefore, it presents a high resistance to the flow of fault current. This resistance is usually enough to restrict fault current to a

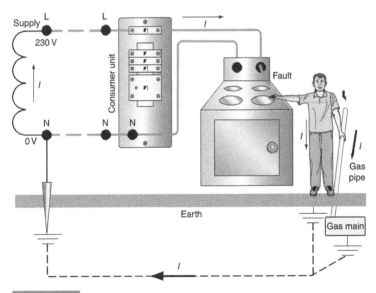

FIGURE A1.3 Shock path.

level well below that of the rating of the protective device, leaving a faulty circuit uninterrupted. Clearly, this is an unhealthy situation.

In all but the most rural areas, consumers can connect to a metallic earth return conductor, which is ultimately connected to the earthed neutral of the supply. This, of course, presents a low-resistance path for fault currents to operate the protection.

In summary, connecting metalwork to earth places that metal at or near zero potential and bonding between metallic parts puts such parts at a similar potential even under fault conditions. Add to this a low-resistance earth fault return path, which will enable the circuit protection to operate very fast, and we have significantly reduced the risk of electric shock.

We can see from this how important it is to check that equipment earthing is satisfactory and that there is no damage to conductor insulation.

SAFE ISOLATION OF SUPPLIES

Before any work is undertaken on low-voltage (50–1000 V AC) installations, supplies should be isolated and proved dead; the procedure is as follows:

1. Identify the circuit or item to be worked on.
2. Switch off/isolate and lock off or place warning notices if locking is not available.
3. Select a suitable approved voltage indicator and check that it works, on a known supply.
4. Test that the circuit or equipment is dead using the tester.
5. Recheck the tester on the known supply again.

Never assume or take someone else's word that supplies are dead and safe to work on. Always check for yourself.

Basic Electrical Theory

This section has been added as a refresher for those electrical operatives who once knew but have now forgotten the basics.

ELECTRICAL QUANTITIES AND UNITS

Quantity	Symbol	Units
Current	I	Ampere (A)
Voltage	V	Volt (V)
Resistance	R	Ohm (Ω)
Power	P	Watt (W)

Current

This is the flow of electrons in a conductor.

Voltage

This is the electrical pressure causing the current to flow.

Resistance

This is the opposition to the flow of current in a conductor determined by its length, cross-sectional area and temperature.

Power

This is the product of current and voltage, hence $P = I \times V$.

Relationship between voltage, current and resistance

Voltage = Current × Resistance $V = I \times R$,
Current = Voltage/Resistance $I = V/R$ or
Resistance = Voltage/Current $R = V/I$

18th Edition IET Wiring Regulations: Wiring Systems and Fault Finding for Installation Electricians. 978-1-138-60611-1.

Common multiples of units

Current (lamperes)	kA	mA
	Kilo-amperes	Milli-amperes
	1000 amperes	1/1000 of an ampere
Voltage V (volts)	kV	mV
	Kilovolts	Millivolts
	1000 volts	1/1000 of a volt
Resistance R (ohms)	MΩ	mΩ
	Megohms	Milli-ohms
	1 000 000 ohms	1/1000 of an ohm
Power P (watts)	MW	kW
	Megawatt	Kilowatt
	1 000 000 watts	1000 watts

Resistance in series

These are resistances joined end to end in the form of a chain. The total resistance increases as more resistances are added (Figure A2.1).

Hence, if a cable length is increased, its resistance will increase in proportion. For example, a 100 m length of conductor has twice the resistance of a 50 m length of the same diameter.

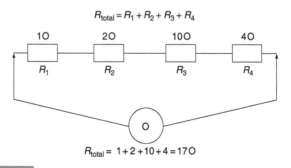

$$R_{total} = R_1 + R_2 + R_3 + R_4$$

1O	2O	10O	4O
R_1	R_2	R_3	R_4

O

$$R_{total} = 1 + 2 + 10 + 4 = 17O$$

FIGURE A2.1   Resistances in series.

Resistance in parallel

These are resistances joined like the rungs of a ladder. Here the total resistance decreases the more there are (Figure A2.2).

The insulation between conductors is in fact countless millions of very high value resistances in parallel. Hence an increase in cable length results in a decrease in insulation resistance. This value is measured in millions of ohms (i.e. megohms, $M\Omega$).

The overall resistance of two or more conductors will also decrease if they are connected in parallel (Figure A2.3).

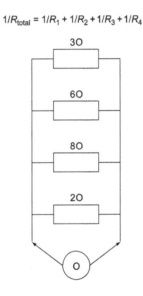

$$1/R_{total} = 1/R_1 + 1/R_2 + 1/R_3 + 1/R_4$$

$$\begin{aligned}
1/R_{total} &= 1/R_1 + 1/R_2 + 1/R_3 + 1/R_4 \\
&= 1/3 + 1/6 + 1/8 + 1/2 \\
&= 0.333 + 0.167 + 0.125 + 0.5 \\
&= 1.125 \\
\therefore R_{total} &= 1/1.125 \\
&= 0.89\ \Omega
\end{aligned}$$

FIGURE A2.2   Resistances in parallel.

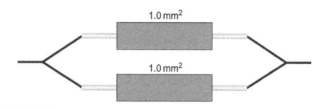

FIGURE A2.3   Conductors in parallel.

The total resistance will be half of either one and would be the same as the resistance of a 2 mm² conductor. Hence resistance decreases if conductor cross-sectional area increases.

Example A2.1

If the resistance of a 1.0 mm² conductor is 19.5 mΩ/m, what would be the resistance of:

1. 85 m of 1.0 mm² conductor
2. 1 m of 6.0 mm² conductor
3. 25 m of 4.0 mm² conductor
4. 12 m of 0.75 mm² conductor.

Answers

1. 1.0 mm² is 19.5 mΩ/m, so, 85 m would be 19.5 × 85/1000 = 1.65 Ω
2. A 6.0 mm² conductor would have a resistance 6 times less than a 1.0 mm² conductor, i.e. 19.5/6 = 3.25 mΩ
3. 25 m of 4.0 mm² would be 19.5 × 25/4 × 1000 = 0.12 Ω
4. 19.5 mΩ/m = 1.5 (the ratio of 0.75 mm² to 1.00 mm² conductor) × 12 m = 0.351 Ω.

POWER, CURRENT AND VOLTAGE

As we have already seen, at a basic level, power = current × voltage, or $P = I \times V$. However, two other formulae can be produced: $P = I^2 \times R$ and $P = V^2/R$. Here are some examples of how these may be used.

1. A 3 kW 230 V immersion heater has ceased to work although fuses, etc., are all intact. A test using a low-resistance ohmmeter should reveal the heaters resistance, which can be determined from:

$P = V^2 / P$

So, $R = V^2/P$

$$= \frac{230 \times 230}{3000} = \frac{52900}{3000} = 17.6 \ \Omega$$

This can be compared with the manufacturer's intended resistance. This would show that the element is not broken and further investigation should take place (probably a faulty thermostat).

2. Two lighting points have been wired, incorrectly, in series. The effect on the light output from two 100 W/230 V lamps connected to these points can be shown as follows:

Each lamp will have a resistance of $R = V^2/P$ (when hot)

$$= \frac{230 \times 230}{100} = \frac{52900}{100} = 529 \ \Omega$$

It will be seen that each lamp will have only 115 V as a supply (Figure A2.4). Hence each will deliver a power of $P = V^2/R$, giving

$$= \frac{115 \times 115}{529} = 25 \ W$$

which is a quarter of its rated value, and so both lamps will be only a quarter of their intended brightness.

3. The current flowing in a 10 m length of 2.5 mm² twin cable is 12 A. The resistance of such cable is approximately 0.015 Ω/m, so the power consumed by the cable would be:

$P = I^2 \times R$

$= 12 \times 12 \times 0.015 \times 10 = 21.6 \ W$

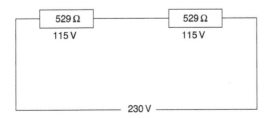

Lamps in series.

Solutions

QUIZ CONTROLLER (CHAPTER 1)

Figure A3.1 shows a solution. Any contestant pushing his/her button energizes their corresponding relay, which is held on via contacts RA1, RB1 or RC1. Two sets of N/C contacts, located in each of the other contestants' circuits, will open, rendering those circuits inoperative. The system is returned to normal when the reset button is pushed, de-energizing the held-on relay.

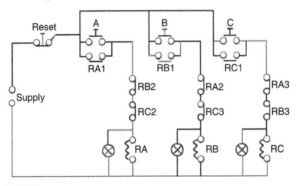

FIGURE A3.1

FIGURE A3.2 Exercise 1a.

FIGURE A3.3 Exercise 1b.

FIGURE A3.4   Exercise 1c.

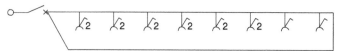

FIGURE A3.5   Exercise 1d.

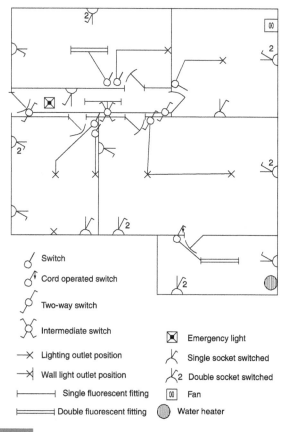

⊸ Switch	
⊸ Cord operated switch	
⊸ Two-way switch	
⋈ Intermediate switch	
⟶✕ Lighting outlet position	⊠ Emergency light
⟶⊣ Wall light outlet position	⋏ Single socket switched
⊢────⊣ Single fluorescent fitting	⋏2 Double socket switched
⊨════⊨ Double fluorescent fitting	⊞ Fan
	⦷ Water heater

FIGURE A3.6   Exercise 2.

Index

Page numbers in *italics* denote an illustration, **bold** indicates a table